Mastering MongoDB 6.x
Third Edition

Expert techniques to run high-volume and fault-tolerant
database solutions using MongoDB 6.x

Alex Giamas

BIRMINGHAM—MUMBAI

Mastering MongoDB 6.x

Third Edition

Copyright © 2022 Packt Publishing

Publishing Product Manager: Heramb Bhavsar

Senior Editor: Tazeen Shaikh

Content Development Editor: Joseph Sunil

Technical Editor: Sweety Pagaria

Copy Editor: Safis Editing

Project Coordinator: Farheen Fathima

Proofreader: Safis Editing

Indexer: Pratik Shirodkar

Production Designer: Alishon Mendonca

Marketing Coordinator: Nivedita Singh

First published: November 2017

Second edition: March 2019

Third Edition: September 2022

Production reference: 2250822

Published by Packt Publishing Ltd.

Livery Place

35 Livery Street

Birmingham

B3 2PB, UK.

ISBN 978-1-80324-386-3

www.packt.com

In memory of my dearest mother Evi (1954-2020) and my highly esteemed father-in-law, Christos (1950-2022) who gave me the guidance and strength to carry on. You will always be in my mind, in my heart and live forever through our memories.

To my son Dimitris, with the wish that he will grow up in a better world than ours. I will always have your back, stand next to you and be your greatest fan. Never doubt my love and only be the best that you can, for yourself and your loved ones. Make the world a better place, please.

Contributors

About the author

Alex Giamas is a freelance consultant and a hands-on Lead Technical and Data Architect. Over the past 15 years, he has expertise in designing and developing systems for the UK Government (HMRC, Cabinet Office, DIT) and private sector (Amazon ProServe, PwC, Fintech Fortune 500, Yahoo!, Verizon) clients. Alex is an alumnus of the MassChallenge London cohort as the co-founder and CTO of a digital health startup. Alex has authored Mastering MongoDB 3.x and 4.x, both by Packt Publishing. Alex has developed large-scale robust, distributed software systems in Python, JavaScript, Ruby, and Java. He is a MongoDB Certified Developer, a Cloudera Hadoop Certified Developer with Data Science Essentials, and a Carnegie Mellon and Stanford graduate.

I would like to thank my wife Mary for her support, patience, and understanding all throughout the journey of writing 3 books on MongoDB in the past 5 years. You are the architect of our life together; you always support me by modeling our non-relational and unstructured daily routine data. I wouldn't have made it without you by my side.

I would like to thank the team at Packt Publishing for their support and understanding when major life and death events got in the way. You people rock!

About the reviewers

Amit Phaltankar is a software developer and a blogger experienced in building lightweight and efficient software components. He specializes in wiring web-based applications and handling large-scale data sets using traditional SQL, NoSQL, and big data technologies. He is experienced in many technology stacks and loves learning and adapting to newer technology trends. Amit is passionate about improving his skill set and loves guiding and grooming his peers and contributing to blogs. He is also an author of MongoDB Fundamentals.

Kevin Smith is a Microsoft MVP and has been working with MongoDB since the early releases in 2010, with the first deployment being a 16-sharded cluster. He has been a technology enthusiast from a young age and enjoys working on a wide range of technologies. In his day-to-day work he is focused on, but not limited to, .NET and TypeScript, and using AWS and Azure Cloud Services. He is heavily involved in the community, running three community groups, 2 in the North of England, dotnet York and dotnetsheff, and one virtual hackathon group, MiniHack. He is passionate about helping and sharing knowledge with others, speaking at user groups and conferences, and contributing to the open source community.

Table of Contents

Part 2 – Querying Effectively

3

MongoDB CRUD Operations

4

Auditing

5

Advanced Querying

6

Multi-Document ACID Transactions

7

Aggregation

8

Indexing

Part 3 – Administration and Data Management

9

Monitoring, Backup, and Security

10

Managing Storage Engines

11

MongoDB Tooling

12

Harnessing Big Data with MongoDB

Part 4 – Scaling and High Availability

13

Mastering Replication

14

Mastering Sharding

15

Fault Tolerance and High Availability

Index

Other Books You May Enjoy

Preface

MongoDB is the leading non-relational database. This book covers all the major features of MongoDB including the latest version, 6. MongoDB 6.x adds many new features and expands on existing ones, such as aggregation, indexing, replication, sharding, and MongoDB Atlas tools. Some of the MongoDB Atlas tools that you will master include Atlas dedicated clusters and Serverless, Atlas Search, Charts, Realm Application Services/Sync, Compass, Cloud Manager, and Data Lake.

Learning from experience and demonstrating code using realistic use cases, you will master the art of modeling, shaping, and querying your data and become the MongoDB oracle for the business. You will dive deep into broadly used as well as niche areas such as optimizing queries, configuring large-scale clusters, configuring your cluster for high performance and availability, and many more. With this under your belt, you will be proficient in auditing, monitoring, and securing your clusters using a structured and organized approach.

By the end of the book, you will have grasped all the practical understanding needed to design, develop, administer and scale MongoDB-based database applications both on-premises and in the cloud.

Who this book is for

The book is geared towards MongoDB developers and database administrators who wish to learn in depth how to model their data using MongoDB, for both greenfield and existing projects. Some understanding of MongoDB, shell command skills, and basic database design concepts is required to get the most out of the book.

What this book covers

Chapter 1, MongoDB – A Database for the Modern Web, will act as a quick refresher of the structure and its key components for businesses. You will learn how the database has evolved over time and how different designs get to be driven by data modeling.

Chapter 2, Schema Design and Data Modeling, will explain the pros and cons of each approach and help you identify the best route to take for each case (key-value, document-based, graph, and CAP theorem). You will learn how to model your data for different use cases along with the tradeoffs of the different designs. Furthermore, you will learn how to configure the drivers for each language to make sure that you are making the most of MongoDB.

Chapter 3, MongoDB CRUD Operations, will showcase the MongoDB shell and its capabilities. This chapter will show you how to perform all the CRUD operations and administration tasks using the shell. You will learn to use the aggregation framework for prototyping, getting quick insights from data and where it shines compared to the older MapReduce framework. You will also learn how to use the new shell, mongosh, and how to migrate from the old one. Finally, you will know how to use the versioned API and the rapid, regular MongoDB release cycle to ensure code sustainability.

Chapter 4, Auditing, explores what auditing is and how is it different from regular application logging. You will learn how to set up auditing on-premises and in the cloud and how to identify irregular activity and single it out. Finally, you will have a case study bringing it all together that can serve as a reference for an end-to-end auditing implementation.

Chapter 5, Advanced Querying, will teach you how to query MongoDB, both using ODM and the driver from Ruby, along with instructions to use from the PHP and Python perspectives. It will show you how to avoid expensive operations and design queries so that they require the least possible maintenance down the line. You will be able to update and delete documents without impacting the underlying storage, and perform complex queries using regular expressions and arrays. You will also learn how and when you need to change streams.

Chapter 6, Multi-Document ACID Transactions, explores the theory behind transactions. How do the different transaction levels compare? When you should use transactions and what are the drawbacks? You will also configure different concern levels for transactions, and find out what the limitations of multi-document transactions in MongoDB are as of version 6.x.

Chapter 7, Aggregation, acts as a deep dive into the aggregation framework and how it can be a solution instead of complex queries or building data pipelines for ETL in code. You will learn how to use aggregation for a generic use case, as well as for more specific cases, such as window operators and time series. You will also be able to create and update materialized views from an RDBMS perspective.

Chapter 8, Indexing, showcases the different types of indexes and how to use them to improve querying efficiency. You will also learn to troubleshoot slow queries and optimize them. You will also understand what the drawbacks are of using too many indexes.

Chapter 9, Monitoring, Backup, and Security, explores monitoring, backups, and deployment for developers. The chapter shows how a developer or DBA can administer a MongoDB server or servers in their own data center or the cloud using MongoDB Atlas. It also shows how you can monitor ongoing operations and keep an eye on the cluster's health, and how can you comply with GDPR using security updates, new in 6.x. Finally, you will learn how to find the optimal path for deployment and upgrading existing clusters.

Chapter 10, Managing Storage Engines, introduces the concept of storage engines. This chapter explains why they matter and how WiredTiger can help users administer MongoDB better.

Chapter 11, MongoDB Tooling, provides an answer to the following questions: how does tooling in the MongoDB ecosystem work? How can you use Realm, Search, Serverless, and Charts to build applications more quickly and robustly? How can you deploy and administer Atlas Kubernetes Operator to manage resources in Atlas without leaving Kubernetes? How can you use data from multiple devices, including IoT and mobile? How is MongoDB innovative in these use cases?

Chapter 12, Harnessing Big Data with MongoDB, showcases the integration of MongoDB with other data sources in the big data ecosystem, HDFS, message queues, and Kafka. It tells you how you can design MongoDB and where your source of truth and aggregator operations should lie when you store and process datasets. You will also learn to use MongoDB Atlas Data Lake as a warehouse and when should you use it as opposed to AWS Data Lake solutions.

Chapter 13, Mastering Replication, explores replication for MongoDB, along with setting up and administering replica sets. You will learn why we need replication as a concept and which workloads require replication as a first-class concern. You will learn to connect and administer replica sets from different drivers. Finally, you will see the latest updates to replica set setup and administration in MongoDB 6.x.

Chapter 14, Mastering Sharding, shows how to horizontally scale a MongoDB installation and how to set up and administer a sharded cluster. You will also learn how to decide on a sharding strategy and reshard data if the requirements change.

Chapter 15, Fault Tolerance and High Availability, explores various tips and tricks that you can use with all the concepts that you covered in the previous chapters.

To get the most out of this book

Software/hardware covered in the book	Operating system requirements
MongoDB	Windows, macOS, or Linux

If you are using the digital version of this book, we advise you to type the code yourself or access the code from the book's GitHub repository (a link is available in the next section). Doing so will help you avoid any potential errors related to the copying and pasting of code.

The code in this book has been tested on Windows and is running as intended. In case of any issues, please raise it in the GitHub repository.

Download the example code files

You can download the example code files for this book from GitHub at `https://github.com/PacktPublishing/Mastering-MongoDB-6.x`. If there's an update to the code, it will be updated in the GitHub repository.

We also have other code bundles from our rich catalog of books and videos available at `https://github.com/PacktPublishing/`. Check them out!

Download the color images

We also provide a PDF file that has color images of the screenshots and diagrams used in this book. You can download it here: `https://packt.link/k275B`.

Conventions used

There are a number of text conventions used throughout this book.

`Code in text`: Indicates code words in text, database table names, folder names, filenames, file extensions, pathnames, dummy URLs, user input, and Twitter handles. Here is an example: "Fields such as `ts` (timestamp), `users`, and `rules` are included in every audit log by default."

A block of code is set as follows:

```
apiVersion: v1
kind: ConfigMap
metadata:
name:<<any sample name we choose(1)>>
namespace: mongodb
data:
projectId:<<Project ID from above>>
baseUrl: <<BaseURI from above>>
```

Any command-line input or output is written as follows:

```
mongod --auditFilter '{ atype: "authenticate", "param.db":
"test" }'
```

Bold: Indicates a new term, an important word, or words that you see onscreen. For instance, words in menus or dialog boxes appear in **bold**. Here is an example: "The project ID is typically a long hex string, for example, `"620173c921b1ab3de3e8e610"`, which we can retrieve from the **Organization | Projects** page."

> **Tips or important notes**
> Appear like this.

Get in touch

Feedback from our readers is always welcome.

General feedback: If you have questions about any aspect of this book, email us at customercare@packtpub.com and mention the book title in the subject of your message.

Errata: Although we have taken every care to ensure the accuracy of our content, mistakes do happen. If you have found a mistake in this book, we would be grateful if you would report this to us. Please visit www.packtpub.com/support/errata and fill in the form.

Piracy: If you come across any illegal copies of our works in any form on the internet, we would be grateful if you would provide us with the location address or website name. Please contact us at copyright@packt.com with a link to the material.

If you are interested in becoming an author: If there is a topic that you have expertise in and you are interested in either writing or contributing to a book, please visit authors.packtpub.com.

Share Your Thoughts

Once you've read *Mastering MongoDB 6.x, Third Edition*, we'd love to hear your thoughts! Scan the QR code below to go straight to the Amazon review page for this book and share your feedback.

https://packt.link/r/1-803-24386-4

Your review is important to us and the tech community and will help us make sure we're delivering excellent quality content.

Part 1 – Basic MongoDB – Design Goals and Architecture

This part of the book covers all the basics surrounding MongoDB and will serve as an introduction to the world of MongoDB. We will go through the history of databases and how the need for MongoDB grew out of database evolution. We will learn how to design our database and model our data for efficiency, ease of development, and maintainability.

This part contains the following chapters:

- *Chapter 1, MongoDB - A Database for the Modern Web*
- *Chapter 2, Schema Design and Data Modeling*

MongoDB – A Database for the Modern Web

In this chapter, we will lay the foundations for understanding MongoDB. We will explore how it is a database designed for the modern web and beyond. Learning is as important as knowing how to learn in the first place. We will go through the references that have the most up-to-date information about MongoDB, for both new and experienced users.

By the end of this chapter, you will have learned where MongoDB is best suited to be used and when it might be sub-optimal to use it. Learning about the evolution of MongoDB and the wider ecosystem will allow you to apply critical thinking when evaluating different database options early on in the software development life cycle.

In this chapter, we will cover the following topics:

- SQL and MongoDB's history and evolution
- MongoDB from the perspective of SQL and other NoSQL technology users
- MongoDB's common use cases and why they matter
- MongoDB's configuration and best practices

Technical requirements

To sail smoothly through the chapter, you will need MongoDB version 5 installed or a free tier account in MongoDB Atlas The code that has been used for all of the chapters in this book can be found at `https://github.com/PacktPublishing/Mastering-MongoDB-6.x`.

The evolution of SQL and NoSQL

Structured Query Language (**SQL**) existed even before the **World Wide Web** (**WWW**). Dr. E. F. Codd originally published a paper, *A Relational Model of Data for Large Shared Data Banks*, in June 1970, in the **Association of Computer Machinery** (**ACM**) journal, **Communications of the ACM**. SQL was initially developed at IBM by Chamberlin and Boyce, in 1974. Relational Software (now known as Oracle Corporation) was the first to develop a commercially available implementation of SQL, which was targeted at United States governmental agencies.

The first **American National Standards Institute** (**ANSI**) SQL standard came out in 1986. Since then, there have been eight revisions, with the most recent being published in 2016 (SQL:2016).

SQL was not particularly popular at the start of the WWW. Static content could just be hardcoded onto the HTML page without much fuss. However, as the functionality of websites grew, webmasters wanted to generate web page content driven by offline data sources, in order to generate content that could change over time without redeploying code.

Common Gateway Interface (**CGI**) scripts, developing Perl or Unix shells, were driving early database-driven websites in Web 1.0. With Web 2.0, the web evolved from directly injecting SQL results into the browser to using two-tier and three-tier architectures that separated views from the business and model logic, allowing for SQL queries to be modular and isolated from the rest of the web application.

On the other hand, **Not only SQL** (**NoSQL**) is much more modern and supervened web evolution, rising at the same time as Web 2.0 technologies. The term was first coined by Carlo Strozzi, in 1998, for his open source database that did not follow the SQL standard but was still relational.

This is not what we currently expect from a NoSQL database. Johan Oskarsson, a developer at Last.fm, reintroduced the term in early 2009, in order to group a set of distributed, non-relational data stores that were being developed. Many of them were based on Google's **Bigtable** and **MapReduce** papers or Amazon's **DynamoDB**, which is a highly available key-value-based storage system.

NoSQL's foundations grew upon relaxed **atomicity, consistency, isolation, and durability** (**ACID**) properties, which guarantee performance, scalability, flexibility, and reduced complexity. Most NoSQL databases have gone one way or the other in providing as many of the previously mentioned qualities as possible, even offering adjustable guarantees to the developer. The following diagram describes the evolution of SQL and NoSQL:

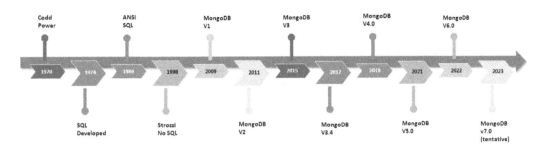

Figure 1.1 – Database evolution

In the next section, we will learn more about how MongoDB has evolved over time, from a basic object store to a full-fledged general-purpose database system.

The evolution of MongoDB

MongoDB Inc's former name, 10gen Inc., started to develop a cloud computing stack in 2007 and soon realized that the most important innovation was centered around the document-oriented database that they built to power it, which was MongoDB. MongoDB shifted from a **Platform as a Service (PaaS)** to an open source model and released MongoDB version 1.0 on August 27, 2009.

Version 1 of MongoDB was pretty basic in terms of features, authorization, and ACID guarantees, but it made up for these shortcomings with performance and flexibility.

In the following sections, we will highlight the major features of MongoDB, along with the version numbers with which they were introduced.

The major feature set for versions 1.0 and 1.2

The major new features of versions 1.0 and 1.2 are listed as follows:

- A document-based model
- A global lock (process level)
- Indexes on collections
- CRUD operations on documents
- No authentication (authentication was handled at the server level)

- Primary and secondary replication: Back then, they were named master and slave, respectively, and were changed to their current names with the SERVER-20608 ticket, in version 4.9.0

- MapReduce (introduced in v1.2)

- Stored JavaScript functions (introduced in v1.2)

Version 2

The major new features of version 2 are listed as follows:

- Background index creation (since v1.4)

- Sharding (since v1.6)

- More query operators (since v1.6)

- Journaling (since v1.8)

- Sparse and covered indexes (since v1.8)

- Compact commands to reduce disk usage

- More efficient memory usage

- Concurrency improvements

- Index performance enhancements

- Replica sets are now more configurable and data center-aware

- MapReduce improvements

- Authentication (since 2.0, for sharding and most database commands)

- Geospatial features introduced

- The aggregation framework (since v2.2) and enhancements (since v2.6)

- **Time-to-Live** (**TTL**) collections (since v2.2)

- Concurrency improvements, among which there is DB-level locking (since v2.2)

- Text searching (since v2.4) and integration (since v2.6)

- Hashed indexes (since v2.4)

- Security enhancements and role-based access (since v2.4)

- A V8 JavaScript engine instead of SpiderMonkey (since v2.4)

- Query engine improvements (since v2.6)

- A pluggable storage engine API

- A WiredTiger storage engine has been introduced, with document-level locking, while the previous storage engine (now called **MMAPv1**) supports collection-level locking

Version 3

The major new features of version 3 are listed as follows:

- Replication and sharding enhancements (since v3.2)

- Document validation (since v3.2)

- The aggregation framework's enhanced operations (since v3.2)

- Multiple storage engines (since v3.2, only in Enterprise Edition)

- Query language and indexes collation (since v3.4)

- Read-only database views (since v3.4)

- Linearizable read concerns (since v3.4)

Version 4

The major new features of version 4 are listed as follows:

- Multi-document ACID transactions (since v4.0)

- Change streams (since v4.0)

- MongoDB tools (Stitch, Mobile, Sync, and Kubernetes Operator) (since v4.0)

- Retryable writes (since v4.0)

- Distributed transactions (since v4.2)

- Removing the outdated MMAPv1 storage engine (since v4.2)

- Updating the shard key (since v4.2)

- On-demand materialized views using aggregation pipelines (since v4.2)

- Wildcard indexes (since v4.2)

- Streaming replication in replica sets (since v4.4)

- Hidden indexes (since v4.4)

Version 5

The major new features of version 5 are listed as follows:

- A quarterly MongoDB release schedule going forward
- Window operators using aggregation pipelines (since v5.0)
- A new MongoDB shell – mongosh (since v5.0)
- Native time series collections (since v5.0)
- Live resharding (since v5.0)
- Versioned APIs (since v5.0)
- Multi-cloud client-side field level encryption (since v5.0)
- Cross-Shard Joins and Graph Traversals (since v5.1)

The following diagram shows MongoDB's evolution over time:

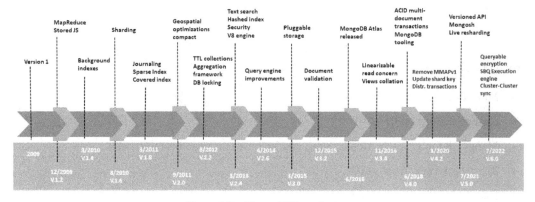

Figure 1.2 – MongoDB's evolution

As you can see, version 1 was pretty basic, whereas version 2 introduced most of the features present in the current version, such as sharding, usable and spatial indexes, geospatial features, and memory and concurrency improvements.

On the way from version 2 to version 3, the aggregation framework was introduced, mainly as a supplement to the aging MapReduce framework that didn't keep up to speed with dedicated frameworks, such as Hadoop. Then, text search was added, and slowly but surely, the performance, stability, and security of the framework improved, adapting to the increasing enterprise loads of customers using MongoDB.

With WiredTiger's introduction in version 3, locking became much less of an issue for MongoDB, as it was brought down from the process (global lock) to the document level, which is almost the most granular level possible.

Version 4 marked a major transition, bridging the SQL and NoSQL world with the introduction of multi-document ACID transactions. This allowed for a wider range of applications to use MongoDB, especially applications that require a strong real-time consistency guarantee. Further, the introduction of change streams allowed for a faster time to market for real-time applications using MongoDB. Additionally, a series of tools have been introduced to facilitate serverless, mobile, and **Internet of Things (IoT)** development.

With version 5, MongoDB is now a **cloud-first database**, with MongoDB Atlas offering full customer support for all major and minor releases going forward. In comparison, non-cloud users only get official support for major releases (for example, version 5 and then version 6). This is complemented by the newly released versioned API approach, which futureproofs applications. Live resharding addresses the major risk of choosing the wrong sharding key, whereas native time series collections and cross-shard lookups using $lookup and $graphlookup greatly improve analytics capabilities and unlock new use cases. End-to-end encryption and multi-cloud support can help implement systems in industries that have unique regulatory needs and also avoid vendor locking. The new mongosh shell is a major improvement over the legacy mongo shell.

Version 6 brings many incremental improvements. Now time series collections support sharding, compression, an extended range of secondary indexes, and updates and deletes (with limitations), making them useful for production use. The new slot-based query execution engine can be used in eligible queries such as $group and $lookup, improving execution time by optimizing query calculations. Finally, queryable encryption and cluster-to-cluster syncing improve the operational and management aspects of MongoDB.

In its current state, MongoDB is a database that can handle heterogeneous workloads ranging from startup **Minimum Viable Product (MVP)** and **Proof of Concept (PoC)** to enterprise applications with hundreds of servers.

MongoDB for SQL developers

MongoDB was developed in the Web 2.0 era. By then, most developers were using SQL or **object-relational mapping (ORM)** tools from their language of choice to access RDBMS data. As such, these developers needed an easy way to get acquainted with MongoDB from their relational background.

Thankfully, there have been several attempts at making SQL-to-MongoDB cheat sheets that explain the MongoDB terminology in SQL terms.

On a higher level, we have the following:

- Databases and indexes (SQL databases)

- Collections (SQL tables)

- Documents (SQL rows)

- Fields (SQL columns)

- Embedded and linked documents (SQL joins)

Further examples of common operations in SQL and their equivalents in MongoDB are shown in the following table:

SQL	MongoDB
Database	Database
Table	Collection
Index	Index
Row	Document
Column	Field
Joins	Embed in a document or link via DBRef
CREATE TABLE employee (name VARCHAR(100))	db.createCollection("employee")
INSERT INTO employees VALUES (Alex, 36)	db.employees.insert({name: "Alex", age: 36})
SELECT * FROM employees	db.employees.find()
SELECT * FROM employees LIMIT 1	db.employees.findOne()
SELECT DISTINCT name FROM employees	db.employees.distinct("name")
UPDATE employees SET age = 37 WHERE name = 'Alex'	db.employees.update({name: "Alex"}, {$set: {age: 37}}, {multi: true})
DELETE FROM employees WHERE name = 'Alex'	db.employees.remove({name: "Alex"})
CREATE INDEX ON employees (name ASC)	db.employees.ensureIndex({name: 1})

Table 1.1 – Common operations in SQL/MongoDB

A few more examples of common operations can be seen at `https://s3.amazonaws.com/info-mongodb-com/sql_to_mongo.pdf`.

Next, we will check out the features that MongoDB has brought for NoSQL developers.

MongoDB for NoSQL developers

As MongoDB has grown from being a niche database solution to the Swiss Army knife of NoSQL technologies, more developers are also coming to it from a NoSQL background.

Putting the SQL versus NoSQL differences aside, it is the users from the columnar-type databases that face the most challenges. With **Cassandra** and **HBase** being the most popular *column-oriented* database management systems, we will examine the differences between them and how a developer can migrate a system to MongoDB. The different features of MongoDB for NoSQL developers are listed as follows:

- **Flexibility**: MongoDB's notion of documents that can contain sub-documents nested in complex hierarchies is really expressive and flexible. This is similar to the comparison between MongoDB and SQL, with the added benefit that MongoDB can more easily map to plain old objects from any programming language, allowing for easy deployment and maintenance.

- **Flexible query model**: A user can selectively index some parts of each document; query based on attribute values, regular expressions, or ranges; and have as many properties per object as needed by the application layer. Primary and secondary indexes, along with special types of indexes (such as sparse ones), can help greatly with query efficiency. Using a JavaScript shell with MapReduce makes it really easy for most developers (and many data analysts) to quickly take a look at data and get valuable insights.

- **Native aggregation**: The aggregation framework provides an **extract-transform-load** (ETL) pipeline for users to extract and transform data from MongoDB, and either load it in a new format or export it from MongoDB to other data sources. This can also help data analysts and scientists to get the slice of data they need in performing data wrangling along the way.

- **Schema-less model**: This is a result of MongoDB's design philosophy to give applications the power and responsibility to interpret the different properties found in a collection's documents. In contrast to Cassandra's or HBase's schema-based approach, in MongoDB, a developer can store and process dynamically generated attributes.

After learning about the major features MongoDB offers to its users, in the next section, we will learn more about the key characteristics and the most widely deployed use cases.

MongoDB's key characteristics and use cases

In this section, we will analyze MongoDB's characteristics as a database. Understanding the features that MongoDB provides can help developers and architects to evaluate the requirements at hand and how MongoDB can help to fulfill them. Also, we will go over some common use cases from the experience of MongoDB, Inc. that have delivered the best results for its users. Finally, we will uncover some of the most common points of criticism against MongoDB and non-relational databases in general.

Key characteristics

MongoDB has grown to become a general-purpose NoSQL database, offering the best of both the RDBMS and NoSQL worlds. Some of the key characteristics are listed as follows:

- **It is a general-purpose database**: In contrast to other NoSQL databases that are built for specific purposes (for example, graph databases), MongoDB can serve heterogeneous loads and multiple purposes within an application. This became even more true after version 4.0

introduced multi-document ACID transactions, further expanding the use cases in which it can be effectively used.

- **Flexible schema design**: Document-oriented approaches with non-defined attributes that can be modified on the fly is a key contrast between MongoDB and relational databases.

- **It is built with high availability, from the ground up**: In the era of five-nines availability, this has to be a given. Coupled with automatic failover upon detection of a server failure, this can help to achieve high uptime.

- **Feature-rich**: Offering the full range of SQL-equivalent operators, along with features such as MapReduce, aggregation frameworks, TTL and capped collections, and secondary indexing, MongoDB can fit many use cases, no matter how diverse the requirements are.

- **Scalability and load balancing**: It is built to scale, both vertically and (mainly) horizontally. Using sharding, an architect can share a load between different instances and achieve both read and write scalability. Data balancing happens automatically (and transparently to the user) via the shard balancer.

- **Aggregation framework**: Having an ETL framework built into the database means that a developer can perform most of the ETL logic before the data leaves the database, eliminating, in many cases, the need for complex data pipelines.

- **Native replication**: Data will get replicated across a replica set without a complicated setup.

- **Security features**: Both authentication and authorization are taken into account so that an architect can secure their MongoDB instances.

- **JSON (BSON and Binary JSON) objects for storing and transmitting documents**: JSON is widely used across the web for frontend and API communication, and, as such, it is easier when the database is using the same protocol.

- **MapReduce**: Even though the MapReduce engine is not as advanced as it is in dedicated frameworks, nonetheless, it is a great tool for building data pipelines.

- **Querying and geospatial information in 2D and 3D**: This might not be critical for many applications, but if it is for your use case, then it is really convenient to be able to use the same database for geospatial calculations and data storage.

- **Multi-document ACID transactions**: Starting from version 4.0, MongoDB supports ACID transactions across multiple documents.

- **Mature tooling**: The tooling for MongoDB has evolved to support systems around DBaaS to Sync, Mobile, and serverless (Stitch).

Use cases for MongoDB

Since MongoDB is a highly popular NoSQL database, there have been several use cases where it has succeeded in supporting quality applications, with a great delivery time to the market.

Many of its most successful use cases center around the following list of areas:

- The integration of siloed data, providing a single view of them
- IoT
- Mobile applications
- Real-time analytics
- Personalization
- Catalog management
- Content management

All of these success stories share some common characteristics. We will try to break them down in order of relative importance:

- **Schema flexibility** is probably the most important one. Being able to store documents inside a collection that can have different properties can help during both the development phase and when ingesting data from heterogeneous sources that may or may not have the same properties. This is in contrast with an RDBMS, where columns need to be predefined, and having sparse data can be penalized. In MongoDB, this is the norm, and it is a feature that most use cases share. Having the ability to deeply nest attributes into documents and add arrays of values into attributes while also being able to search and index these fields helps application developers to exploit the schema-less nature of MongoDB.

- **Scaling** and **sharding** are the most common patterns for MongoDB use cases. Easily scaling using built-in sharding, using replica sets for data replication, and offloading primary servers from read loads can help developers store data effectively.

- Additionally, many use cases use MongoDB as a way of **archiving data**. Used as a pure data store (and without the need to define schemas), it is fairly easy to dump data into MongoDB to be analyzed at a later date by business analysts, using either the shell or some of the numerous BI tools that can integrate easily with MongoDB. Breaking data down further, based on time caps or document counts, can help serve these datasets from RAM, the use case in which MongoDB is most effective.

- **Capped collections** are also a feature used in many use cases. Capped collections can restrict documents in a collection by count or by the overall size of the collection. In the latter case, we need to have an estimate of the size per document, in order to calculate how many documents will fit into our target size. Capped collections are a quick and dirty solution used to answer requests such as "*Give me the last hour's overview of the logs*" without the need for maintenance and running async background jobs to clean our collection. Oftentimes, they might be used to quickly build and operate a queuing system. Instead of deploying and maintaining a dedicated queuing system, such as ActiveMQ, a developer can use a collection to store messages, and then use the native tailable cursors provided by MongoDB to iterate through the results as they pile up and feed an external system. Alternatively, you can use a TTL index within a regular collection if they require greater flexibility.

- **Low operational overhead** is also a common pattern in many use cases. Developers working in agile teams can operate and maintain clusters of MongoDB servers without the need for a dedicated DBA. The free cloud monitoring service can greatly help in reducing administrative overhead for Community Edition users, whereas MongoDB Atlas, the hosted solution by MongoDB, Inc., means that developers do not need to deal with operational headaches.

- In terms of business sectors using MongoDB, there is a huge variety coming from almost all industries. A common pattern seems to be higher usage where we are more interested in aggregated data than individual transaction-level data. Fields such as IoT can benefit the most by exploiting the availability over consistent design, storing lots of data from sensors in a cost-efficient way. On the other hand, financial services have absolutely stringent consistency requirements, aligned with proper ACID characteristics that make MongoDB more of a challenge to adapt. A financial transaction might be small in size but big in impact, which means that we cannot afford to leave a single message without proper processing.

- Location-based data is also a field where MongoDB has thrived, with Foursquare being one of the most prominent early clients. MongoDB offers quite a rich set of features around two-dimensional and three-dimensional geolocation data, such as searching by distance, geofencing, and intersections between geographical areas.

- Overall, the rich feature set is a common pattern across different use cases. By providing features that can be used in many different industries and applications, MongoDB can be a unified solution for all business needs, offering users the ability to minimize operational overhead and, at the same time, iterate quickly in product development.

MongoDB criticism

MongoDB's criticism can be broken down into the following points:

- MongoDB has had its fair share of criticism throughout the years. The web-scale proposition has been met with skepticism by many developers. The counterargument is that scale is not needed most of the time, and the focus should be on other design considerations. While this might occasionally be true, it is a false dichotomy, and in an ideal world, we would have both. MongoDB is as close as it can get to combining scalability with features, ease of use, and time to market.

- MongoDB's schema-less nature is also a big point of debate and argument. Schema-less can be really beneficial in many use cases, as it allows for heterogeneous data to be dumped into the database without complex cleansing and without ending up with lots of empty columns or blocks of text stuffed into a single column. On the other hand, this is a double-edged sword, as a developer could end up with many documents in a collection that have loose semantics in their fields, and it can become really hard to extract these semantics at the code level. If our schema design is not optimal, we could end up with a data store, rather than a database.

- A lack of proper ACID guarantees is a recurring complaint from the relational world. Indeed, if a developer needs access to more than one document at a time, it is not easy to guarantee RDBMS properties, as there are no transactions. In the RDBMS sense, having no transactions also means that complex writes will need to have application-level logic to roll back. If you need to update three documents in two collections to mark an application-level transaction complete, and the third document does not get updated for whatever reason, the application will need to undo the previous two writes – something that might not exactly be trivial.

- With the introduction of multi-document transactions in version 4, MongoDB can cope with ACID transactions at the expense of speed. While this is not ideal, and transactions are not meant to be used for every CRUD operation in MongoDB, it does address the main source of criticism.

- The configuration setup defaults that favored setting up MongoDB but not operating it in a production environment are disapproved. For years, the default write behavior was **write and forget**; sending a write wouldn't wait for an acknowledgment before attempting the next write, resulting in insane write speeds with poor behaviors in the case of failure. Also, authentication is an afterthought, leaving thousands of MongoDB databases on the public internet prey to whoever wants to read the stored data. Even though these were conscious design decisions, they are decisions that have affected developers' perceptions of MongoDB.

It's important to note that MongoDB has addressed all of the shortcomings throughout the years, with the aim of becoming a versatile and resilient general-purpose database system. Now that we understand the characteristics and features of MongoDB, we will learn how to configure and set up MongoDB efficiently.

MongoDB configuration and best practices

In this section, we will present some of the best practices around operations, schema design, durability, replication, sharding, security, and AWS. Further information on when to implement these best practices will be presented in later chapters.

Operational best practices

As a database, MongoDB is built with developers in mind, and it was developed during the web era, so it does not require as much operational overhead as traditional RDBMS. That being said, there are some best practices that need to be followed to be proactive and achieve high availability goals.

In order of importance, the best practices are as follows:

- **Mind the location of your data files**: Data files can be mounted anywhere by using the --dbpath command-line option. It is really important to ensure that data files are stored in partitions with sufficient disk space, preferably **XFS**, or at the very least **Ext4**.

- **Keep yourself updated with versions**: Before version 5, there was a different versioning naming convention. Even major numbered versions are the stable ones. So, 3.2 is stable, whereas 3.3 is not. In this example, 3.3 is the developmental version that will eventually materialize into the stable 3.4 version. It is a good practice to always update to the latest updated security version (which, at the time of writing this book, is 4.0.2) and to consider updating as soon as the next stable version comes out (4.2, in this example).

  Version 5 has become cloud-first. The newest versions are automatically updated in MongoDB Atlas with the ability to opt out of them, whereas all versions are available to download for evaluation and development purposes. *Chapter 3, MongoDB CRUD Operations*, goes into more detail about the new rapid release schedule and how it affects developers and architects.

- **Use MongoDB Cloud monitoring**: The free MongoDB, Inc. monitoring service is a great tool to get an overview of a MongoDB cluster, notifications, and alerts and to be proactive about potential issues.

- **Scale up if your metrics show heavy use**: Do not wait until it is too late. Utilizing more than 65% in CPU or RAM, or starting to notice disk swapping, should be the threshold to start thinking about scaling, either vertically (by using bigger machines) or horizontally (by sharding).

- **Be careful when sharding**: Sharding is a strong commitment to your shard key. If you make the wrong decision, it might be really difficult to go back from an operational perspective. When designing for sharding, architects need to take deep consideration of the current workloads (reads/writes) and what the current and expected data access patterns are. Live resharding, which was introduced in version 5, mitigates the risk compared to previous versions, but it's still better to spend more time upfront instead of resharding after the fact. Always use the shard key in queries or else MongoDB will have to query all shards in the cluster, negating the major sharding advantage.

- **Use an application driver maintained by the MongoDB team**: These drivers are supported and tend to get updated faster than drivers with no official support. If MongoDB does not support the language that you are using yet, please open a ticket in MongoDB's JIRA tracking system.

- **Schedule regular backups**: No matter whether you are using standalone servers, replica sets, or sharding, a regular backup policy should also be used as a second-level guard against data loss. XFS is a great choice as a filesystem, as it can perform snapshot backups.

- **Manual backups should be avoided**: When possible, regular, automated backups should be used. If we need to resort to a manual backup, we can use a hidden member in a replica set to take the backup from. We have to make sure that we are using `db.fsync` with `{lock: true}` in this member, to get the maximum consistency at this node, along with having journaling turned on. If this volume is on AWS, we can get away with taking an EBS snapshot straight away.

- **Enable database access control**: Never put a database into a production system without access control. Access control should be implemented at a node level, by a proper firewall that only allows access to specific application servers to the database, and at a DB level, by using the built-in roles or defining custom-defined ones. This has to be initialized at start-up time by using the `--auth` command-line parameter and can be configured by using the `admin` collection.

- **Test your deployment using real data**: Since MongoDB is a schema-less, document-oriented database, you might have documents with varying fields. This means that it is even more important than with an RDBMS to test using data that resembles production data as closely as possible. A document with an extra field of an unexpected value can make the difference between an application working smoothly or crashing at runtime. Try to deploy a staging server using production-level data, or at least fake your production data in staging, by using an appropriate library, such as Faker for Ruby.

Schema design best practices

MongoDB is schema-less, and you need to design your collections and indexes to accommodate for this fact:

- **Index early and often**: Identify common query patterns, using cloud monitoring, the GUI that MongoDB Compass offers, or logs. Analyzing the results, you should create indexes that cover the most common query patterns, using as many indexes as possible at the beginning of a project.

- **Eliminate unnecessary indexes**: This is a bit counter-intuitive to the preceding suggestion, but monitor your database for changing query patterns, and drop the indexes that are not being used. An index will consume RAM and I/O, as it needs to be stored and updated along with the documents in the database. Using an aggregation pipeline and `$indexStats`, a developer can identify the indexes that are seldom being used and eliminate them.

- **Use a compound index, rather than index intersection**: Most of the time, querying with multiple predicates (*A* and *B*, *C* or *D* and *E*, and so on) will work better with a single compound index than with multiple simple indexes. Also, a compound index will have its data ordered by field, and we can use this to our advantage when querying. An index on fields *A*, *B*, and *C* will be used in queries for *A*, *(A,B)*, *(A,B,C)*, but not in querying for *(B,C)* or *(C)*.

- **Low selectivity indexes**: Indexing a field on gender, for example, will statistically return half of our documents back, whereas an index on the last name will only return a handful of documents with the same last name.

- **Use regular expressions**: Again, since indexes are ordered by value, searching using a regular expression with leading wildcards (that is, `/.*BASE/`) won't be able to use the index. Searching with trailing wildcards (that is, `/DATA.*/`) can be efficient, as long as there are enough case-sensitive characters in the expression.

- **Avoid negation in queries**: Indexes are indexing values, not the absence of them. Using `NOT` in queries, instead of using the index, can result in full table scans.

- **Use partial indexes**: If we need to index a subset of the documents in a collection, partial indexes can help us to minimize the index set and improve performance. A partial index will include a condition on the filter that we use in the desired query.

- **Use document validation**: Use document validation to monitor for new attributes being inserted into your documents and decide what to do with them. With document validation set to warn, we can keep a log of documents that were inserted with new, never-seen-before attributes that we did not expect during the design phase and decide whether we need to update our index or not.

- **Use MongoDB Compass**: MongoDB's free visualization tool is great for getting a quick overview of our data and how it grows over time.

- **Respect the maximum document size of 16 MB**: The maximum document size for MongoDB is 16 MB. This is a fairly generous limit, but it is one that should not be violated under any circumstances. Allowing for documents to grow unbounded should not be an option, and, as efficient as it might be to embed documents, we should always keep in mind that this should be kept under control. Additionally, we should keep track of the average and maximum document sizes using monitoring or the `bsonSize()` method and the aggregation operator.

- **Use the appropriate storage engine**: MongoDB has introduced several new storage engines since version 3.2. The in-memory storage engine should be used for real-time workloads, whereas the encrypted storage engine (only available in MongoDB Enterprise Edition) should be the engine of choice when there are strict requirements around data security. Otherwise, the default WiredTiger storage engine is the best option for general-purpose workloads.

Examining some schema design best practices, we will move on to the best practices for write durability as of MongoDB version 6.

Best practices for write durability

Write durability can be fine-tuned in MongoDB, and, according to our application design, it should be as strict as possible, without affecting our performance goals.

Fine-tune the data and flush it to the disk interval in the WiredTiger storage engine; the default is to flush data to the disk every 60 seconds after the last checkpoint. This can be changed by using the `--wiredTigerCheckpointDelaySecs` command-line option.

MongoDB version 5 has changed the default settings for read and write concerns.

The default write concern is now `majority` writes, which means that in a replica set of three nodes (with one primary and two secondaries), the operation returns as soon as two of the nodes acknowledge it by writing it to the disk. Writes always go to the primary and then get propagated asynchronously to the secondaries. In this way, MongoDB eliminates the possibility of data rollback in the event of a node failure.

If we use arbiters in our replica set, then writes will still be acknowledged solely by the primary if the following formula resolves to true:

#arbiters > #nodes*0.5 - 1

For example, in a replica set of three nodes of which one is the arbiter and two are storing data, this formula resolves to the following:

1 > 3*0.5 - 1 ... 1 > 0.5 ... true

> **Note**
> MongoDB 6 restricts the number of arbiters to a maximum of one.

The default read concern is now `local` instead of `available`, which mitigates the risk of returning orphaned documents for reads in sharded collections. Orphaned documents might be returned during chunk migrations, which can be triggered either by MongoDB or, since version 5, also by the user when applying live resharding to the sharded collection.

Multi-document ACID transactions and the transactional guarantees that they have provided since MongoDB 4.2, coupled with the introduction of streaming replication and *replicate-before-journaling* behavior, have improved replication performance. Additionally, they allow for more durable and consistent default write and read concerns without affecting performance as much. The new defaults are promoting durability and consistent reads and should be carefully evaluated before changing them.

Best practices for replication

Under the right conditions, replica sets are MongoDB's mechanism to provide redundancy, high availability, and higher read throughput. In MongoDB, replication is easy to configure and focuses on operational terms:

- **Always use replica sets**: Even if your dataset is currently small, and you don't expect it to grow exponentially, you never know when that might happen. Also, having a replica set of at least three servers helps to design for redundancy, separating the workloads between real time and analytics (using the secondary) and having data redundancy built-in from day one. Finally, there are some corner cases that you will identify earlier by using a replica set instead of a single standalone server, even for development purposes.

- **Use a replica set to your advantage**: A replica set is not just for data replication. We can (and, in most cases, should) use the primary server for writes and preference reads from one of the secondaries to offload the primary server. This can be done by setting read preferences for reads, along with the correct write concern, to ensure that writes propagate as needed.

- **Use an odd number of replicas in a MongoDB replica set**: If a server is down or loses connectivity with the rest of them (network partitioning), the rest have to vote as to which one will be elected as the primary server. If we have an odd number of replica set members, we can guarantee that each subset of servers knows if they belong to the majority of the minority of the replica set members. If we cannot have an odd number of replicas, we need to have one extra host set as an arbiter, with the sole purpose of voting in the election process. Even a micro-instance in EC2 could serve this purpose.

Best practices for sharding

Sharding is MongoDB's solution for horizontal scaling. In *Chapter 9, Monitoring, Backup, and Security*, we will go over its usage in more detail, but the following list offers some best practices, based on the underlying data architecture:

- **Think about query routing**: Based on different shard keys and techniques, the mongos query router might direct the query to some (or all) of the members of a shard. It is important to take our queries into account when designing sharding. This is so that we don't end up with our queries hitting all of our shards.

- **Use tag-aware sharding**: Tags can provide more fine-grained distribution of data across our shards. Using the right set of tags for each shard, we can ensure that subsets of data get stored in a specific set of shards. This can be useful for data proximity between application servers, MongoDB shards, and the users.

Best practices for security

Security is always a multi-layered approach, and the following recommendations do not form an exhaustive list; they are just the bare basics that need to be done in any MongoDB database:

- Always turn authentication on. There are multiple hacks over the years where open MongoDB servers have been hacked for fun or profit such as being backed up and deleted to extort admins to pay. It is a good practice to set up authentication even in non-production environments to decrease the possibility of human error.

- The HTTP status interface should be disabled.

- The RESTful API should be disabled.

- The JSON API should be disabled.

- Connect to MongoDB using SSL.

- Audit the system activity.

- Use a dedicated system user to access MongoDB with appropriate system-level access.

- Disable server-side scripting if it is not needed. This will affect MapReduce, built-in `db.group()` commands, and `$where` operations. If they are not used in your code base, it is better to disable server-side scripting at startup by using the `--noscripting` parameter or setting `security.javascriptEnabled` to false in the configuration file.

After examining the best practices for security in general, we will dive into what are the best practices for AWS deployments.

Best practices for AWS

When we are using MongoDB, we can use our own servers in a data center, a MongoDB-hosted solution such as MongoDB Atlas, or we can rent instances from Amazon by using EC2. EC2 instances are virtualized and share resources in a transparent way, with collocated VMs in the same physical host. So, there are some more considerations to take into account if you wish to go down that route, as follows:

- Use EBS-optimized EC2 instances.

- Get EBS volumes with provisioned **I/O operations per second** (**IOPS**) for consistent performance.

- Use EBS snapshotting for backup and restore.

- Use different availability zones for high availability and different regions for disaster recovery. Using different availability zones within each region that Amazon provides guarantees that our data will be highly available. Different regions should mostly be used for disaster recovery in case a catastrophic event ever takes out an entire region. A region might be EU-West-2 (for London), whereas an availability zone is a subdivision within a region; currently, three availability zones are available in the London region.

- Deploy globally, access locally.

- For truly global applications with users from different time zones, we should have application servers in different regions access the data that is closest to them, using the right read preference configuration in each server.

Reference documentation and further reading

Reading a book is great (and reading this book is even better), but continuous learning is the only way to keep up to date with MongoDB.

The online documentation available at `https://docs.mongodb.com/manual/` is the perfect starting point for every developer, new or seasoned.

The JIRA tracker is a great place to take a look at fixed bugs and the features that are coming up next: `https://jira.mongodb.org/browse/SERVER/`.

Some other great books on MongoDB are listed as follows:

- *MongoDB Fundamentals*: A hands-on guide to using MongoDB and Atlas in the real world, by Amit Phaltankar and Juned Ahsan
- *MongoDB: The Definitive Guide 3e: Powerful and Scalable Data Storage*, by Shannon Bradshaw and Eoin Brazil
- *MongoDB Topology Design: Scalability, Security, and Compliance on a Global Scale*, by Nicholas Cottrell
- Any book by Kristina Chodorow

The MongoDB user group (`https://groups.google.com/forum/#!forum/mongodb-user`) has a great archive of user questions about features and long-standing bugs. It is a place to go when something doesn't work as expected.

Online forums (such as Stack Overflow and Reddit, among others) are always a source of knowledge, with the caveat that something might have been posted a few years ago and might not apply anymore. Always check before trying.

Finally, MongoDB University is a great place to keep your skills up to date and to learn about the latest features and additions: `https://university.mongodb.com/`.

Summary

In this chapter, we started our journey through web, SQL, and NoSQL technologies, from their inception to their current states. We identified how MongoDB has been shaping the world of NoSQL databases over the years, and how it is positioned against other SQL and NoSQL solutions.

We explored MongoDB's key characteristics and how MongoDB has been used in production deployments. We identified the best practices for designing, deploying, and operating MongoDB.

Initially, we identified how to learn by going through the documentation and online resources that can be used to stay up-to-date with the latest features and developments.

In the next chapter, we will go deeper into schema design and data modeling. We will look at how to connect to MongoDB by using both the official drivers and an **Object Document Mapper** (**ODM**), which is a variation of object-relational mappers for NoSQL databases.

2

Schema Design and Data Modeling

This chapter will focus on schema design for schemaless databases such as MongoDB. Although this may sound counterintuitive, there are considerations that we should take into account when we develop for MongoDB. We will learn about the schema considerations and the data types supported by MongoDB. We will also learn about preparing data for text searches in MongoDB by connecting using Ruby, Python, and PHP.

By the end of this chapter, you will have learned how to connect to MongoDB using the Ruby, Python, and PHP languages and using the low-level driver commands or an object-relational mapping framework. You will have learned how to model your data for different entity relationship mappings and what the trade-offs of different design decisions are.

In this chapter, we will cover the following topics:

- Relational schema design
- Data modeling
- Modeling data for atomic operations
- Modeling relationships
- Connecting to MongoDB

Technical requirements

To follow along with the code in this chapter, you need to install MongoDB locally or connect to a MongoDB Atlas database using a UNIX type system. You can download the MongoDB Community Edition from mongodb.com or use the fully managed DBaaS MongoDB Atlas offering, which provides a free tier as well as seamless upgrades to the latest version.

You will also need to download the official drivers for the language of your choice – Ruby, Python, or PHP. Finally, you will need to download MongoId, PyMODM, or Doctrine ODM frameworks for the respective languages. You can find all the code from this chapter in the GitHub repository at https://github.com/PacktPublishing/Mastering-MongoDB-6.x.

Relational schema design

In relational databases, we design with the goal of avoiding data anomalies and data redundancy.

Data anomalies can happen when we have the same information stored in multiple columns; we update one of them but not the rest and so end up with conflicting information for the same column of information.

Another example of a data anomaly is when we cannot delete a row without losing the information that we need, possibly in other rows referenced by it.

Data redundancy on the other hand refers to a situation where our data is not in a normal form but has duplicate data across different tables. This can lead to data inconsistency and make the data integrity difficult to maintain.

In relational databases, we use normal forms to normalize our data. Starting from the basic **first normal form** (**1NF**), onto the **second normal form** (**2NF**), **third normal form** (**3NF**), and **Boyce-Codd normal form** (**BCNF**), there are different ways to model our data. We structure our tables and columns by using functional dependencies between different data units. Following this formal method, we can prove that our data is normalized, with the downside that we can sometimes end up with many more tables than the domain model objects that we originally started with from our systems model.

In practice, relational database modeling is often driven by the structure of the data that we have. In web applications following some sort of **model-view-controller** (**MVC**) model pattern, we will model our database according to our models that are based on the **Unified Modeling Language** (**UML**) diagram conventions. Abstractions such as the ORM for **Django** or the **Active Record** for Rails help application developers abstract database structure to object models. Ultimately, many times, we end up designing our database based on the structure of the available data. Therefore, we are designing around the questions that we need to answer.

MongoDB schema design

In contrast to relational databases, in MongoDB, we have to base our modeling on our application-specific data access patterns. Finding out the questions that our users will have is paramount to designing our entities. In contrast to an RDBMS, data duplication and denormalization are used far more frequently, for good reason.

The document model that MongoDB uses means that every document can hold substantially more or less information than the next one, even within the same collection. Coupled with rich and detailed queries being possible in MongoDB at the embedded document level, this means that we are free to design our documents in any way that we want. When we know our data access patterns, we can estimate which fields need to be embedded and which can be split out into different collections.

Read-write ratio

The read-write ratio is often an important design consideration for MongoDB modeling. When reading data, we want to avoid scatter-gather situations, where we have to hit several shards with random I/O requests to get the data that our application needs.

When writing data, on the other hand, we want to spread out writes to as many servers as possible to avoid overloading any single one of them. These goals appear to be conflicting on the surface, but they can be combined once we know our access patterns, coupled with application design considerations, such as using a replica set to read from secondary nodes.

Data modeling

In this section, we will discuss the different types of data MongoDB uses, how they map to the data types that programming languages use, and how we can model data relationships in MongoDB using Ruby, Python, and PHP.

Data types

MongoDB uses BSON, a binary-encoded serialization for JSON documents. BSON extends the JSON data types, offering, for example, native data and binary data types.

BSON, compared to protocol buffers, allows for more flexible schemas, which comes at the cost of space efficiency. In general, BSON is space-efficient, easy to traverse, and time-efficient in encoding/decoding operations, as can be seen in the following table (see the MongoDB documentation at `https://docs.mongodb.com/manual/reference/bson-types/`):

Type	Number	Alias	Notes
Double	1	`double`	
String	2	`string`	
Object	3	`object`	
Array	4	`array`	
Binary data	5	`binData`	
ObjectID	7	`objectId`	
Boolean	8	`bool`	
Date	9	`date`	
Null	10	`null`	
Regular expression	11	`regex`	
JavaScript	13	`javascript`	
32-bit integer	16	`int`	
Timestamp	17	`timestamp`	
64-bit integer	18	`long`	
Decimal128	19	`decimal`	New in version 3.4
Min. key	-1	`minKey`	
Max. key	127	`maxKey`	
Undefined	6	`undefined`	Deprecated
DBPointer	12	`dbPointer`	Deprecated
Symbol	14	`symbol`	Deprecated
JavaScript (with scope)	15	`javascriptWithScope`	Deprecated

Table 2.1 – MongoDB data types

In MongoDB, we can have documents with different value types for a given field and we can distinguish between them when querying using the $type operator.

For example, if we have a balance field in GBP with 32-bit integers and double data types, and based on whether the balance field has in it or not, we can easily query by $type for all accounts that have a rounded balance field, with any of the following queries shown in the example:

```
db.account.find( { "balance" : { $type : 16 } } );
db.account.find( { "balance" : { $type : "integer" } } );
```

We will compare the different data types in the following section.

Comparing different data types

Due to the nature of MongoDB, it's perfectly acceptable to have different data type objects in the same field. This may happen by accident or on purpose (that is, null and actual values in a field).

The sorting order of different types of data, from highest to lowest, is as follows:

1. Max. key (internal type)
2. Regular expression
3. Timestamp
4. Date
5. Boolean
6. ObjectID
7. Binary data
8. Array
9. Object
10. Symbol, string
11. Numbers (`int`, `long`, `double`, `Decimal128`)
12. Null
13. Min. key (internal type)

Non-existent fields get sorted as if they have `null` in the respective field. Comparing arrays is a bit more complex than fields. Ascending order of comparison (or <) will compare the smallest element of each array. Descending order of comparison (or >) will compare the largest element of each array.

For example, see the following scenario:

```
> db.types.find()
{ "_id" : ObjectId("5908d58455454e2de6519c49"), "a" : [ 1, 2, 3
] }
{ "_id" : ObjectId("5908d59d55454e2de6519c4a"), "a" : [ 2, 5 ]
}
```

In ascending order, this is as follows:

```
> db.types.find().sort({a:1})
{ "_id" : ObjectId("5908d58455454e2de6519c49"), "a" : [ 1, 2, 3
] }
{ "_id" : ObjectId("5908d59d55454e2de6519c4a"), "a" : [ 2, 5 ]
}
```

However, in descending order, it is as follows:

```
> db.types.find().sort({a:-1})
{ "_id" : ObjectId("5908d59d55454e2de6519c4a"), "a" : [ 2, 5 ]
}
{ "_id" : ObjectId("5908d58455454e2de6519c49"), "a" : [ 1, 2, 3
] }
```

The same applies when comparing an array with a single number value, as illustrated in the following example. Inserting a new document with an integer value of 4 is done as follows:

```
> db.types.insert({"a":4})
WriteResult({ "nInserted" : 1 })
```

The following example shows the code snippet for a descending sort:

```
> db.types.find().sort({a:-1})
{ "_id" : ObjectId("5908d59d55454e2de6519c4a"), "a" : [ 2, 5 ]
}
{ "_id" : ObjectId("5908d73c55454e2de6519c4c"), "a" : 4 }
{ "_id" : ObjectId("5908d58455454e2de6519c49"), "a" : [ 1, 2, 3
] }
```

And the following example is the code snippet for an ascending sort:

```
> db.types.find().sort({a:1})
{ "_id" : ObjectId("5908d58455454e2de6519c49"), "a" : [ 1, 2, 3
] }
{ "_id" : ObjectId("5908d59d55454e2de6519c4a"), "a" : [ 2, 5 ]
}
{ "_id" : ObjectId("5908d73c55454e2de6519c4c"), "a" : 4 }
```

In each case, we have highlighted the values being compared in bold.

We will learn about the date type in the following section.

Date types

Dates are stored as milliseconds, with effect from January 01, 1970 (epoch time). They are 64-bit signed integers, allowing for a range of 135 million years before and after 1970. A negative date value denotes a date before January 01, 1970. The BSON specification refers to the date type as UTC DateTime.

Dates in MongoDB are stored in UTC. There isn't a `timestamp` field with a `timezone` data type like in some relational databases. Applications that need to access and modify timestamps based on local time should store the `timezone` offset together with the date and offset dates on an application level.

In the MongoDB shell, this could be done using the following format with JavaScript:

```
var now = new Date();
db.page_views.save({date: now,
                    offset: now.getTimezoneOffset()});
```

Then, you need to apply the saved offset to reconstruct the original local time, as in the following example:

```
var record = db.page_views.findOne();
var localNow = new Date( record.date.getTime() - ( record.
offset * 60000 ) );
```

In the next section, we will cover `ObjectId`.

ObjectId

`ObjectId` is a special data type for MongoDB. Every document has an `_id` field from cradle to grave. It is the primary key for each document in a collection and has to be unique. If we omit this field in a `create` statement, it will be assigned automatically with an `ObjectId` data type.

Messing with `ObjectId` is not advisable but we can use it (with caution!) for our purposes.

`ObjectId` has the following distinctions:

- It has 12 bytes
- It is ordered
- Sorting by `_id` will sort by creation time for each document, down to one-second granularity
- The creation time can be accessed by `.getTimeStamp()` in the shell

The structure of an `ObjectId` value consists of the following:

- The first four bytes are the seconds since the Unix epoch, 00:00:00 UTC on January 01, 1970.
- The next five bytes are unique to the device and process. They are generated randomly with the first three bytes being the machine identifier and the next two the process ID.
- The last three bytes represent an incrementing counter, starting with a random value.

The following diagram shows the structure of an `ObjectId` value:

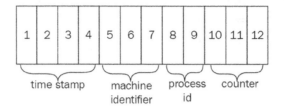

Figure 2.1 – Internal structure of the ObjectId value

By design, `ObjectId` will be unique across different documents in replica sets and sharded collections.

In the next section, we will learn about modeling data for atomic operations.

Modeling data for atomic operations

MongoDB is relaxing many of the typical **Atomicity, Consistency, Isolation, and Durability (ACID)** constraints found in RDBMS. The default operation mode does not support transactions, making it important to keep the state consistent across operations, especially in the event of failures.

Some operations are atomic at the document operation level:

- `update()`
- `findandmodify()`
- `remove()`

These are all atomic (all-or-nothing) for a single document.

This means that, if we embed information in the same document, we can make sure they are always in sync.

An example would be an inventory application, with a document per item in our inventory. Every time a product is placed in a user's shopping cart, we decrement the `available_now` value by one and append the `userid` value to the `shopping_cart_by` array.

With `total_available = 5`, `available_now = 3`, and `shopping_cart_count = 2`, this use case could look like the following:

```
{available_now : 3, shopping_cart_by: ["userA", "userB"] }
```

When someone places the item in their shopping cart, we can issue an atomic update, adding their user ID in the `shopping_cart_by` field and, at the same time, decreasing the `available_now` field by one.

This operation will be guaranteed to be atomic at the document level. If we need to update multiple documents within the same collection, the update operation may complete successfully without modifying all of the documents that we intended it to. This could happen because the operation is not guaranteed to be atomic across multiple document updates.

This pattern can help in some but not all cases. In many cases, we need multiple updates to be applied on all or nothing across documents, or even collections.

A typical example would be a bank transfer between two accounts. We want to subtract x GBP from user A, then add x to user B. If we fail to do either of these two steps, both balances would return to their original state.

Since version 4, we should use multi-document transactions in such cases, which we will cover in *Chapter 6, Multi-Document ACID Transactions*.

In the next section, we will learn more about the visibility of MongoDB operations between multiple readers and writers.

Read isolation and consistency

MongoDB read operations would be characterized as *read uncommitted* in a traditional RDBMS definition. What this means is that, by default, reads can get values that may not finally persist to the disk in the event of, for example, data loss or a replica set rollback operation.

In particular, when updating multiple documents with the default write behavior, lack of isolation may result in the following:

- Reads may miss documents that were updated during the update operations
- Non-serializable operations
- Read operations are not point-in-time

Queries with cursors that don't use `.snapshot()` may also, in some cases, get inconsistent results. This can happen if the query's resultant cursor fetches a document that receives an update while the query is still fetching results, and, because of insufficient padding, ends up in a different physical location on the disk, ahead of the query's result cursor position. `.snapshot()` is a solution for this edge case, with the following limitations:

- It doesn't work with sharding
- It doesn't work with `sort()` or `hint()` to force an index to be used
- It still won't provide point-in-time read behavior

If our collection has mostly static data, we can use a unique index in the query field to simulate `snapshot()` and still be able to apply `sort()` to it.

All in all, we need to apply safeguards at the application level to make sure that we won't end up with unexpected results.

Starting from version 3.4, MongoDB offers linearizable read concern. With linearizable read concern from the primary member of a replica set and a majority write concern, we can ensure that multiple threads can read and write a single document as if a single thread were performing these operations one after the other. This is considered a linearizable schedule in RDBMS, and MongoDB calls it the real-time order.

Starting from version 4.4, we can set a global default read concern on the replica set and sharded cluster level. The implicit write concern is `w:majority`, which means that the write will be acknowledged after it's been propagated to a majority of the nodes in the cluster.

If we use a read concern that is *majority* or higher, then we can make sure that the data we read is write-committed in the majority of nodes, making sure that we avoid the *read uncommitted* problem that we described at the beginning of this section.

Modeling relationships

In the following sections, we will explain how we can translate relationships in RDBMS theory into MongoDB's document collection hierarchy. We will also examine how we can model our data for text search in MongoDB.

One-to-one

Coming from the relational DB world, we identify objects by their relationships. A one-to-one relationship could be a person with an address. Modeling it in a relational database would most probably require two tables: a **person** and an **address** table with a `person_id` foreign key in the **address** table, as shown in the following diagram:

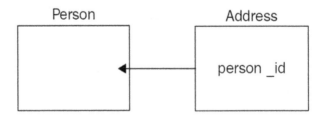

Figure 2.2 – Foreign key used to model a one-to-one relationship in MongoDB

The perfect analogy in MongoDB would be two collections, `Person` and `Address`, as shown in the following code:

```
> db.Person.findOne()
{
"_id" : ObjectId("590a530e3e37d79acac26a41"), "name" : "alex"
}
> db.Address.findOne()
{
"_id" : ObjectId("590a537f3e37d79acac26a42"),
"person_id" : ObjectId("590a530e3e37d79acac26a41"),
"address" : "N29DD"
}
```

Now, we can use the same pattern as we do in a relational database to find `Person` from `address`, as shown in the following example:

```
> db.Person.find({"_id": db.Address.
findOne({"address":"N29DD"}).person_id})
{
"_id" : ObjectId("590a530e3e37d79acac26a41"), "name" : "alex"
}
```

This pattern is well known and works well in the relational world.

> **Note**
> The command is performing two queries nested inside one another, the first one to retrieve the `person_id` `_id` value, which we use to query the `Person` collection by `_id`.

In MongoDB, we don't have to follow this pattern, as there are more suitable ways to model this kind of relationship.

One way in which we would typically model one-to-one or one-to-many relationships in MongoDB would be through embedding. If the person has two addresses, then the same example would then be shown in the following way:

```
{ "_id" : ObjectId("590a55863e37d79acac26a43"), "name" :
"alex", "address" : [ "N29DD", "SW1E5ND" ] }
```

Using an embedded array, we can access every `address` that this user has. Embedding querying is rich and flexible so that we can store more information in each document, as shown in the following example:

```
{ "_id" : ObjectId("590a56743e37d79acac26a44"),
"name" : "alex",
"address" : [ { "description" : "home", "postcode" : "N29DD" },
{ "description" : "work", "postcode" : "SW1E5ND" } ] }
```

The advantages of this approach are as follows:

- No need for two queries across different collections

- It can exploit atomic updates to make sure that updates in the document will be all-or-nothing from the perspective of other readers of this document

- It can embed attributes in multiple nest levels, creating complex structures

The most notable disadvantage is that the maximum size of the document is 16 MB, so this approach cannot be used for an arbitrary, ever-growing number of attributes. Storing hundreds of elements in embedded arrays will also degrade performance.

One-to-many and many-to-many

When the number of elements on the *many* side of the relationship can grow unbounded, it's better to use references. References can come in two forms:

1. From the *one* side of the relationship, store an array of many-sided elements, as shown in the following example:

   ```
   > db.Person.findOne()
   { "_id" : ObjectId("590a530e3e37d79acac26a41"), "name" :
   "alex", addresses:
   [ ObjectID('590a56743e37d79acac26a44'),
   ObjectID('590a56743e37d79acac26a46'),
   ObjectID('590a56743e37d79acac26a54') ] }
   ```

2. This way, we can get the array of the `addresses` elements from the *one* side and then query with `in` to get all the documents from the *many* side, as shown in the following example:

   ```
   > person = db.Person.findOne({"name":"mary"})
   > addresses = db.Addresses.find({_id: {$in: person.
   addresses} })
   ```

Turning this one-to-many into many-to-many is as easy as storing this array at both ends of the relationship (that is, in the `Person` and `Address` collections).

3. From the *many* side of the relationship, store a reference to the *one* side, as shown in the following example:

```
> db.Address.find()
{ "_id" : ObjectId("590a55863e37d79acac26a44"),
"person":  ObjectId("590a530e3e37d79acac26a41"),
"address" : [ "N29DD" ] }
{ "_id" : ObjectId("590a55863e37d79acac26a46"),
"person":  ObjectId("590a530e3e37d79acac26a41"),
"address" : [ "SW1E5ND" ] }
{ "_id" : ObjectId("590a55863e37d79acac26a54"),
"person":  ObjectId("590a530e3e37d79acac26a41"),
"address" : [ "N225QG" ] }
> person = db.Person.findOne({"name":"alex"})
> addresses = db.Addresses.find({"person": person._id})
```

As we can see, with both designs we need to make two queries to the database to fetch the information. The second approach has the advantage that it won't let any document grow unbounded, so it can be used in cases where one-to-many is one-to-millions.

Modeling data for keyword searches

Searching for keywords in a document is a common operation for many applications. We can search using an exact match or by using a $regex regular expression in the content of a field that contains text. MongoDB also provides the ability to search using an array of keywords.

The basic need for a keyword search is to be able to search the entire document for keywords. For example, there could be a need to search a document in the products collection, as shown in the following code:

```
{ name : "Macbook Pro late 2016 15in" ,
  manufacturer : "Apple" ,
  price: 2000 ,
  keywords : [ "Macbook Pro late 2016 15in", "2000", "Apple",
"macbook", "laptop", "computer" ]
 }
```

We can create a multikey index in the keywords field, as shown in the following code:

```
> db.products.createIndex( { keywords: 1 } )
```

Now, we can search in the `keywords` field for any name, manufacturer, price, and also any of the custom keywords that we set up. This is not an efficient or flexible approach, as we need to keep keywords lists in sync, we can't use stemming, and we can't rank results (it's more like filtering than searching). The only advantage of this method is that it is slightly quicker to implement.

A better way to solve this problem is by using the special text index type, now in version 3.

Only one `text` index per collection (except for Atlas Search SaaS) can be declared in one or multiple fields. The `text` index supports stemming, tokenization, exact phrase (" "), negation (-), and weighting results.

Index declaration on three fields with custom `weights` is shown in the following example:

```
db.products.createIndex({
    name: "text",
    manufacturer: "text",
    price: "text"
    },
    {
    weights: { name: 10,
        manufacturer: 5,
        price: 1 },
    name: "ProductIndex"
    })
```

In this example, `name` is `10` times more important than `price` but only two times more important than `manufacturer`.

A `text` index can also be declared with a wildcard, matching all the fields that match the pattern, as shown in the following example:

```
db.collection.createIndex( { "$**": "text" } )
```

This can be useful when we have unstructured data and we may not know all the fields that it will come with. We can drop the index by name, just like with any other index.

The greatest advantage though, other than all these features, is that all record keeping is done by the database.

Modeling data for Internet of Things

Internet of things (IoT) is one of the most quickly growing industries and this comes with unique challenges around data storage and processing. IoT systems typically use multiple sensors to gather data that needs to be stored, analyzed, and processed in near real time.

MongoDB introduced **time series collections** in version 5.0 and greatly extended supported functionality in version 6. These are special collections that are faster for data series that can contain sensor measurements.

For example, to create a time series collection, stocks with a `timeField` field of "`timestamp`", and using the default granularity of "`seconds`" (or "`minutes`", "`hours`", et cetera) would require the following mongo shell command:

```
db.createCollection(
    "stocks",
    {
        timeseries: {
            timeField: "timestamp",
            metaField: "metadata",
            granularity: "seconds",
        },
            expireAfterSeconds: 3600,
    }
)
```

`metaField` is a field that can store any kind of metadata that is useful for our querying, such as sensor-unique IDs.

`expireAfterSeconds` is optional and we can set it to allow MongoDB to auto-delete collection data after the threshold.

The time series collection has an index on `timeField`, so we can query it really effectively within any time period that we need. For more complex queries, we can use the **aggregation framework**.

In the next section, we will learn how to connect to MongoDB.

Connecting to MongoDB

There are two ways to connect to MongoDB. The first is by using the driver for your programming language. The second is by using an ODM layer to map your model objects to MongoDB in a transparent way. In this section, we will cover both ways, using three of the most popular languages for web application development: Ruby, Python, and PHP.

Connecting using Ruby

Ruby was one of the first languages to have support from MongoDB with an official driver. The official MongoDB Ruby driver on GitHub is the recommended way to connect to a MongoDB instance. Perform the following steps to connect MongoDB using Ruby:

1. Installation is as simple as adding it to the `Gemfile`, as shown in the following example:

    ```
    gem 'mongo', '~> 2.17'
    ```

2. Then, in our class, we can connect to a database, as shown in the following example:

    ```
    require 'mongo'
    client = Mongo::Client.new([ '127.0.0.1:27017' ],
    database: 'test')
    ```

3. This is the simplest example possible: connecting to a single database instance called `test` in our `localhost`. In most use cases, we would at least have a replica set to connect to, as shown in the following snippet:

    ```
    client_host = ['server1_hostname:server1_ip, server2_
    hostname:server2_ip']
      client_options = {
        database: 'YOUR_DATABASE_NAME',
        replica_set: 'REPLICA_SET_NAME',
        user: 'YOUR_USERNAME',
        password: 'YOUR_PASSWORD'
        }
    client = Mongo::Client.new(client_host, client_options)
    ```

4. The `client_host` servers are seeding the client driver with servers to attempt to connect. Once connected, the driver will determine the server that it has to connect to according to the primary/secondary read or write configuration. The `replica_set` attribute needs to match `REPLICA_SET_NAME` to be able to connect.

5. `user` and `password` are optional but highly recommended in any MongoDB instance. It's good practice to enable authentication by default in the `mongod.conf` file and we will learn more about this in *Chapter 8, Indexing*.

6. Connecting to a sharded cluster is similar to a replica set, with the only difference being that, instead of supplying the server host/port, we need to connect to the MongoDB process that serves as the MongoDB router.

> **Note**
>
> You need to install Ruby, then install RVM from `https://rvm.io/rvm/install`, and finally, run `gem install bundler` for this.

After learning how to connect using the low-level Ruby library, we will learn how to use an **Object Document Mapping (ODM)** library in the next section.

Mongoid ODM

Using a low-level driver to connect to the MongoDB database is often not the most efficient route. All the flexibility that a low-level driver provides is offset against longer development times and code to glue our models with the database.

An ODM can be the answer to these problems. Just like ORMs, ODMs bridge the gap between our models and the database. In Rails, the most widely-used MVC framework for Ruby – Mongoid – can be used to model our data in a similar way to Active Record.

Installing `gem` is similar to the Mongo Ruby driver, by adding a single file in the Gemfile, as shown in the following code:

```
gem 'mongoid', '~> 7.4'
```

Depending on the version of Rails, we may need to add the following to `application.rb` as well:

```
config.generators do |g|
g.orm :mongoid
end
```

Connecting to the database is done through a `mongoid.yml` configuration file. Configuration options are passed as key-value pairs with semantic indentation. Its structure is similar to `database.yml`, used for relational databases.

Some of the options that we can pass through the `mongoid.yml` file are shown in the following table:

Option value	Description
`Database`	The database name.
`Hosts`	Our database hosts.
`Write/w`	The write concern (default is 1).
`Auth_mech`	Authentication mechanism. Valid options are: `:scram`, `:mongodb_cr`, `:mongodb_x509`, and `:plain`. The default option on 3.0 is `:scram`, whereas the default on 2.4 and 2.6 is `:plain`.

Option value	Description
`Auth_source`	The authentication source for our authentication mechanism.
`Min_pool_size/max_pool_size`	Minimum and maximum pool size for connections.
`SSL, ssl_cert, ssl_key, ssl_key_pass_phrase, ssl_verify`	A set of options regarding SSL connections to the database.
`Include_root_in_json`	Includes the root model name in JSON serialization.
`Include_type_for_serialization`	Includes the `_type` field when serializing MongoDB objects.
`Use_activesupport_time_zone`	Uses active support's time zone when converting timestamps between server and client.

Table 2.2 – Mongoid configuration options

The next step is to modify our models to be stored in MongoDB. This is as simple as including one line of code in the model declaration, as shown in the following example:

```
class Person
  include Mongoid::Document
End
```

We can also use the following code:

```
include Mongoid::Timestamps
```

We use it to generate `created_at` and `updated_at` fields in a similar way to `Active Record`. Data fields do not need to be declared by type in our models, but it's good practice to do so. The supported data types are as follows:

- `Array`
- `BigDecimal`
- `Boolean`
- `Date`
- `DateTime`
- `Float`
- `Hash`
- `Integer`
- `BSON::ObjectId`
- `BSON::Binary`
- `Range`

- Regexp
- String
- Symbol
- Time
- TimeWithZone

If the types of fields are not defined, fields will be cast to the object and stored in the database. This is slightly faster but doesn't support all types. If we try to use BigDecimal, Date, DateTime, or Range, we will get back an error.

Inheritance with Mongoid models

The following code is an example of inheritance using the Mongoid models:

```
class Canvas
  include Mongoid::Document
  field :name, type: String
  embeds_many :shapes
end

class Shape
  include Mongoid::Document
  field :x, type: Integer
  field :y, type: Integer
  embedded_in :canvas
end

class Circle < Shape
  field :radius, type: Float
end

class Rectangle < Shape
  field :width, type: Float
  field :height, type: Float
end
```

Now, we have a `Canvas` class with many `Shape` objects embedded in it. `Mongoid` will automatically create a field, which is `_type`, to distinguish between parent and child node fields. When a document is inherited from its fields, relationships, validations, and scopes will propagate down into the child document.

The opposite will not happen; `embeds_many` and `embedded_in` pairs will create embedded subdocuments to store the relationships. If we want to store these via referencing to `ObjectId`, we can do so by substituting these with `has_many` and `belongs_to`.

Connecting using Python

A strong contender to Ruby and Rails is Python and Django. Similar to `Mongoid`, there is MongoEngine and an official MongoDB low-level driver, PyMongo.

Installing PyMongo can be done using `pip` or `easy_install`, as shown in the following code:

```
python -m pip install pymongo
python -m easy_install pymongo
```

Then, in our class, we can connect to a database, as shown in the following example:

```
>>> from pymongo import MongoClient
>>> client = MongoClient()
```

Connecting to a replica set requires a set of seed servers for the client to find out what the primary, secondary, or arbiter nodes in the set are, as indicated in the following example:

```
client = pymongo.MongoClient('mongodb://user:passwd@
node1:p1,node2:p2/?replicaSet=rsname')
```

Using the connection string URL, we can pass a username and password and the `replicaSet` name all in a single string. Some of the most interesting options for the connection string URL are presented in the next section.

Connecting to a shard requires the server host and IP for the MongoDB router, which is the MongoDB process.

PyMODM ODM

Similar to Ruby's Mongoid, PyMODM is an ODM for Python that follows Django's built-in ORM closely. Installing `pymodm` can be done via `pip`, as shown in the following code:

```
pip install pymodm
```

Then, we need to edit `settings.py` and replace the `ENGINE` database with a dummy database, as shown in the following code:

```
DATABASES = {
    'default': {
        'ENGINE': 'django.db.backends.dummy'
    }
}
```

Then we add our connection string anywhere in `settings.py`, as shown in the following code:

```
from pymodm import connect
connect("mongodb://localhost:27017/myDatabase",
alias="MyApplication")
```

Here, we have to use a connection string that has the following structure:

```
mongodb://[username:password@]host1[:port1][,host2[:port2],...
[,hostN[:portN]]][/[database][?options]]
```

Options have to be pairs of `name=value` with an & between each pair. Some interesting pairs are shown in the following table:

Name	Description
minPoolSize/maxPoolSize	Minimum and maximum pool size for connections.
w	Write concern option.
wtimeoutMS	Timeout for write concern operations.
Journal	Journal options.
readPreference	Read preference to be used for replica sets. Available options are: primary, primaryPreferred, secondary, secondaryPreferred, nearest.
maxStalenessSeconds	Specifies in seconds how stale (data lagging behind the primary) a secondary can be before the client stops using it for read operations.
SSL	Using SSL to connect to the database.
authSource	Used in conjunction with username, this specifies the database associated with the user's credentials. When we use external authentication mechanisms, this should be $external for LDAP or Kerberos.
authMechanism	An authentication mechanism can be used for connections. Available options for MongoDB are **SCRAM-SHA-1, MONGODB-CR, MONGODB-X.509**. MongoDB Enterprise (the paid version) offers two more options: **GSSAPI** (Kerberos), **PLAIN (LDAP SASL)**

Table 2.3 – PyMODM configuration options

Model classes need to inherit from `MongoModel`. The following code shows what a sample class will look like:

```
from pymodm import MongoModel, fields
class User(MongoModel):
    email = fields.EmailField(primary_key=True)
    first_name = fields.CharField()
    last_name = fields.CharField()
```

This has a `User` class with `first_name`, `last_name`, and `email` fields, where `email` is the primary field.

Inheritance with PyMODM models

Handling one-to-one and one-to-many relationships in MongoDB can be done using references or embedding. The following example shows both ways, which are references for the model user and embedding for the comment model:

```
from pymodm import EmbeddedMongoModel, MongoModel, fields

class Comment(EmbeddedMongoModel):
    author = fields.ReferenceField(User)
    content = fields.CharField()

class Post(MongoModel):
    title = fields.CharField()
    author = fields.ReferenceField(User)
    revised_on = fields.DateTimeField()
    content = fields.CharField()
    comments = fields.EmbeddedDocumentListField(Comment)
```

Similar to `Mongoid` for Ruby, we can define relationships as being embedded or referenced depending on our design decision.

Connecting using PHP

The MongoDB PHP driver was rewritten from scratch around five years ago to support the PHP 5, PHP 7, and HHVM architectures. The current architecture is shown in the following diagram:

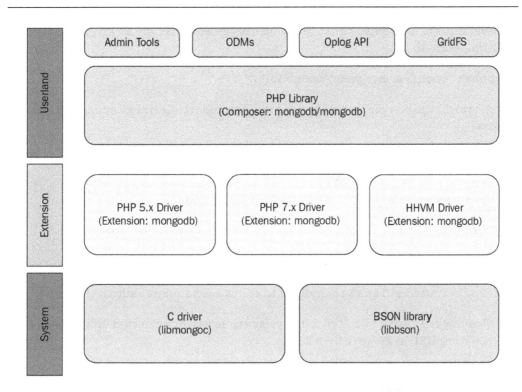

Figure 2.3 – PHP driver architecture

Currently, we have official drivers for all three architectures with full support for the underlying functionality.

Installation is a two-step process. The first step is to install the MongoDB extension. This extension is dependent on the version of PHP (or HHVM) that we have installed and can be done using `brew` in macOS. The following example is with PHP 7.0:

```
brew install php70-mongodb
```

Whereas this example uses PECL:

```
pecl install mongodb
```

Then, copy the following line and place it at the end of your `php.ini` file:

```
extension=mongodb.so
```

The `php -i | grep mongodb` output should then reference `libmongoc` and `libmongocrypt`.

The second step is to use `composer` (a widely-used dependency manager for PHP), as shown in the following example:

```
composer require mongodb/mongodb
```

Connecting to the database can be done by using the connection string URL or by passing an array of options.

Using the connection string URL, we have the following code:

```
$client = new MongoDB\Client($uri = 'mongodb://127.0.0.1/',
array $uriOptions = [], array $driverOptions = [])
```

For example, to connect to a replica set using SSL authentication, we use the following code:

```
$client = new MongoDB\Client('mongodb://
myUsername:myPassword@rs1.example.com,rs2.example.
com/?ssl=true&replicaSet=myReplicaSet&authSource=admin');
```

Alternatively, we can use the `$uriOptions` parameter to pass in parameters without using the connection string URL, as shown in the following code:

```
$client = new MongoDB\Client(
    'mongodb://rs1.example.com,rs2.example.com/'
    [
        'username' => 'myUsername',
        'password' => 'myPassword',
        'ssl' => true,
        'replicaSet' => 'myReplicaSet',
        'authSource' => 'admin',
    ],
);
```

The `$uriOptions` set and the connection string URL options available are analogous to the ones used for Ruby and Python.

Doctrine ODM

Laravel is one of the most widely-used MVC frameworks for PHP, similar in architecture to Django and Rails, from the Python and Ruby worlds respectively. We will follow through with configuring our models using Laravel, Doctrine, and MongoDB. This section assumes that Doctrine is installed and working with Laravel 5.x.

Doctrine entities are **Plain Old PHP Objects** (**POPO**) that, unlike with **Eloquent**, Laravel's default ORM doesn't need to inherit from the Model class. Doctrine uses the **data mapper pattern**, whereas Eloquent uses Active Record. Skipping the get() and set() methods, a simple class would be shown in the following way:

```php
use Doctrine\ORM\Mapping AS ORM;
use Doctrine\Common\Collections\ArrayCollection;
/**
 * @ORM\Entity
 * @ORM\Table(name="scientist")
 */
class Scientist
{
    /**
     * @ORM\Id
     * @ORM\GeneratedValue
     * @ORM\Column(type="integer")
     */
    protected $id;
    /**
     * @ORM\Column(type="string")
     */
    protected $firstname;
    /**
     * @ORM\Column(type="string")
     */
    protected $lastname;
    /**
     * @ORM\OneToMany(targetEntity="Theory",
mappedBy="scientist", cascade={"persist"})
     * @var ArrayCollection|Theory[]
     */
    protected $theories;
    /**
     * @param $firstname
     * @param $lastname
     */
```

```php
    public function __construct($firstname, $lastname)
    {
        $this->firstname = $firstname;
        $this->lastname  = $lastname;
        $this->theories = new ArrayCollection;
    }

...

    public function addTheory(Theory $theory)
    {
        if(!$this->theories->contains($theory)) {
            $theory->setScientist($this);
            $this->theories->add($theory);
        }
    }
}
```

This POPO-based model uses annotations to define field types that need to be persisted in MongoDB. For example, @ORM\Column(type="string") defines a field in MongoDB, with the firstname and lastname string types as the attribute names in the respective lines.

There is a whole set of annotations available here: https://doctrine2.readthedocs.io/en/latest/reference/annotations-reference.html.

If we want to separate the POPO structure from annotations, we can also define them using YAML or XML instead of inlining them with annotations in our POPO model classes.

Inheritance with Doctrine

Modeling one-to-one and one-to-many relationships can be done via annotations, YAML, or XML. Using annotations, we can define multiple embedded subdocuments within our document, as shown in the following example:

```php
/** @Document */
class User
{
    // ...
    /** @EmbedMany(targetDocument="Phonenumber") */
    private $phonenumbers = array();
    // ...
}
/** @EmbeddedDocument */
```

```
class Phonenumber
{
    // ...
}
```

Here, a `User` document embeds many phone numbers. `@EmbedOne()` will embed one subdocument to be used for modeling one-to-one relationships.

Referencing is similar to embedding, as shown in the following example:

```
/** @Document */
class User
{
    // ...
    /**
     * @ReferenceMany(targetDocument="Account")
     */
    private $accounts = array();
    // ...
}
/** @Document */
class Account
{
    // ...
}
```

`@ReferenceMany()` and `@ReferenceOne()` are used to model one-to-many and one-to-one relationships via referencing into a separate collection.

Summary

In this chapter, we have learned about schema design for relational databases and MongoDB and how we can achieve the same goal starting from a different starting point.

In MongoDB, we have to think about read-write ratios, the questions that our users will have in the most common cases, and cardinality among relationships.

We have learned about atomic operations and how we can construct our queries so that we can have ACID properties without the overhead of transactions.

We have also learned about MongoDB data types, how they can be compared, and some special data types, such as `ObjectId`, which can be used both by the database and to our own advantage.

Starting from modeling simple one-to-one relationships, we have gone through one-to-many and also many-to-many relationship modeling, without the need for an intermediate table, as we would do in a relational database, either using references or embedded documents.

We have learned how to model data for keyword searches, one of the features that most applications need to support in a web context.

IoT is a rapidly evolving field and MongoDB provides special support for it. In this chapter, we have learned how to use time series collections to model and store sensor readings.

Finally, we have explored different use cases for using MongoDB with three of the most popular web programming languages. We saw examples using Ruby with the official driver and Mongoid ODM. Then, we explored how to connect using Python with the official driver and PyMODM ODM, and lastly, we worked through an example using PHP with the official driver and Doctrine ODM.

With all these languages (and many others), there are both official drivers offering support and full access functionality to the underlying database operations and also **object data modeling** frameworks, for ease of modeling our data and rapid development.

In the next chapter, we will dive deeper into the MongoDB shell and the operations we can achieve using it. We will also master using the drivers for CRUD operations on our documents.

Part 2 – Querying Effectively

This part of the book goes through basic and advanced querying using the **MongoDB Query Language** (**MQL**). We will learn how to use transactions and the aggregation framework to supplement MQL querying. We will understand how to index our data for more efficient querying and why auditing is important in every production-level database system.

This part contains the following chapters:

3

MongoDB CRUD Operations

In this chapter, we will learn how to use the mongo shell for database administration operations. Starting with simple **create, read, update, and delete** (**CRUD**) operations, we will master scripting from the shell. We will also learn how to write MapReduce scripts from the shell and contrast them to the aggregation framework, into which we will dive deeper in *Chapter 7, Aggregation*. The new mongosh shell, which replaces the legacy mongo shell, is our next learning goal for the chapter. We will learn what the new MongoDB *Stable* (previously named *Versioned*) **application programming interface** (**API**) means for developers and what are the guarantees it provides. Finally, we will explore authentication and authorization using the MongoDB Community Edition and its paid counterpart, the Enterprise Edition.

In this chapter, we will cover the following topics:

- CRUD using the shell
- Administration
- MapReduce in the mongo shell
- Aggregation framework
- Securing the shell
- Authentication with MongoDB
- The new mongosh shell
- MongoDB Stable API

Technical requirements

To follow along with the code in this chapter, you need to install MongoDB locally or connect to a MongoDB Atlas database. You can download the MongoDB Community Edition from mongodb.com or use the fully managed **database-as-a-service** (**DBaaS**) MongoDB Atlas offering, which provides a free tier as well as seamless upgrades to the latest version.

Most examples are compatible with the newest mongosh shell but you may want to use the legacy mongo shell or enable mongosh backwards compatibility using mongosh snippets.

You will also need to download the official drivers for the language of your choice— Ruby, Python, or **PHP: Hypertext Preprocessor (PHP)**. You can find all the code from this chapter in the GitHub repository at `https://github.com/PacktPublishing/Mastering-MongoDB-6.x`.

CRUD using the shell

The mongo shell is equivalent to the administration console used by relational databases. Connecting to the mongo shell is as easy as typing the following code:

```
$ mongosh
```

Type this on the command line for standalone servers or replica sets. Inside the shell, you can view available databases simply by typing the following code:

```
$ db
```

Then, you can connect to a database by typing the following code:

```
> use <database_name>
```

The mongo shell can be used to query and update data in our databases. Inserting this document into the `books` collection can be done like so:

```
> db.books.insertOne({title: 'mastering mongoDB', isbn: '101'})
{
  acknowledged: true,
  insertedIds: { '0': ObjectId("627f8178d95a9c017a390229") }
```

We can then find documents from a collection named `books` by typing the following code:

```
> db.books.find()
{ "_id" : ObjectId("592033f6141daf984112d07c"), "title" :
"mastering mongoDB", "isbn" : "101" }
```

The result we get back from MongoDB informs us that the write succeeded and inserted one new document in the database.

Deleting this document has similar syntax and results in the following code:

```
> db.books.removeOne({isbn: '101'})
  { acknowledged: true, deletedCount: 1 }
```

You can try to update this same document, as shown in the following code block:

```
> db.books.updateOne({isbn:'101'}, {$set: { price: 30}})
{
    acknowledged: true,
    insertedId: null,
    matchedCount: 1,
    modifiedCount: 1,
    upsertedCount: 0
}
> db.books.find()
{ "_id" : ObjectId("592034c7141daf984112d07d"), "price" : 30 }
```

Here, we notice a couple of things, as outlined here:

- The **JavaScript Object Notation (JSON)**-like formatted field in the updateOne command is our query for searching for documents to update

- The WriteResult object notifies us that the query matched one document and modified one document

- Most importantly, the contents of this document were entirely replaced by the contents of the second JSON-like formatted field but we have lost information on title and isbn

By default, the updateOne command in MongoDB will replace the contents of our document with the document we specify in the second argument. If we want to update the document and add new fields to it, we need to use the $set operator, as follows:

```
> db.books.updateOne({isbn:'101'}, {$set: {price: 30}})

{
    acknowledged: true,
    insertedId: null,
    matchedCount: 1,
    modifiedCount: 1,
    upsertedCount: 0
}
```

Now, our document matches what we would expect, as we can see here:

```
> db.books.find()
{ "_id" : ObjectId("592035f6141daf984112d07f"), "title" :
"mastering mongoDB", "isbn" : "101", "price" : 30 }
```

However, deleting a document can be done in several ways. The simplest way is through its unique `ObjectId` value, as illustrated here:

```
> db.books.deleteOne({_id:
ObjectId("592035f6141daf984112d07f")}, justOne: true)
{ acknowledged: true, deletedCount: 1 }
> db.books.find()
>
```

You can see here that when there are no results, the mongo shell will not return anything other than the shell prompt itself: >.

Scripting for the mongo shell

Administering the database using built-in commands is helpful, but it's not the main reason for using the shell. The true power of the mongo shell comes from the fact that it is also a JavaScript shell such that we can perform complex administrative tasks that require a set of commands to execute as one.

We can declare and assign variables in the shell, like so:

```
> var title = 'MongoDB in a nutshell'
> title
MongoDB in a nutshell
> db.books.insertOne({title: title, isbn: 102})

{
  acknowledged: true,
  insertedIds: { '0': ObjectId("627f8178d95a9c017a390229") }
}> db.books.find()
{ "_id" : ObjectId("59203874141daf984112d080"), "title" :
"MongoDB in a nutshell", "isbn" : 102 }
```

In the previous example, we declared a new `title` variable as `MongoDB in a nutshell` and used the variable to insert a new document into our `books` collection, as shown in the previous code snippet.

As it's a JavaScript shell, we can use it for functions and scripts that generate complex results from our database, as illustrated here:

```
> queryBooksByIsbn = function(isbn) { return db.books.
find({isbn: isbn})}
```

With this one-liner, we are creating a new function named `queryBooksByIsbn` that takes a single argument, which is the `isbn` value. With the data that we have in our collection, we can use our new function and fetch books by `isbn`, as shown in the following code snippet:

```
> queryBooksByIsbn("101")
{ "_id" : ObjectId("592035f6141daf984112d07f"), "title" :
"mastering mongoDB", "isbn" : "101", "price" : 30 }
```

Using the shell, we can write and test these scripts. Once we are satisfied, we can store them in the `.js` file and invoke them directly from the command line, like so:

```
$ mongosh <script_name>.js
```

Here are some useful notes about the default behavior of these scripts:

- Write operations will use a default write concern of `w:majority`. Majority write concern is calculated as the minimum value between the calculated majority of all nodes and the number of data-bearing nodes. This means that when we have arbiters that are not data-bearing nodes, they will not be included in the calculation of a majority. For example, in a replica set with three servers with one being an arbiter, we would calculate majority as the minimum value between the "calculated majority of all nodes," which is 2 (the closest number to bring the figure to over 50%) and the number of data-bearing nodes, which is—again—2.

- To get results from operations from a script back to **standard output** (**stdout**), we must use either JavaScript's built-in `print()` function or the Mongo-specific `printjson()` function, which prints out results formatted in JSON.

In the next section, we will examine the difference between scripting for the mongo shell and using it directly.

Differences between scripting for the mongo shell and using it directly

When writing scripts for the mongo shell, we cannot use shell helpers. MongoDB's commands—such as `use <database_name>`, `show collections`, and other helpers—are built into the shell and so are not available from the JavaScript context where our scripts will get executed. Fortunately, there are equivalents to them that are available from the JavaScript execution context, as shown in the following table:

Shell helpers	JavaScript equivalents
`show dbs, show databases`	`db.adminCommand('listDatabases')`
`use <database_name>`	`db = db.getSiblingDB('<database_name>')`
`show collections`	`db.getCollectionNames()`
`show users`	`db.getUsers()`
`show roles`	`db.getRoles({showBuiltinRoles: true})`
`show log <logname>`	`db.adminCommand({ 'getLog' : '<logname>' })`
`show logs`	`db.adminCommand({ 'getLog' : '*' })`

`it`	`cur = db.collection_name_here.find()`
	
	`if ( cur.hasNext() ){`
	
	`  cur.next();`
	
	`}`

Table 3.1 – Helper functions in JavaScript and MongoDB shell

In the previous table, `it` is the iteration cursor that the mongo shell returns when we query and get back too many results to show in one batch.

Using the mongo shell, we can script almost anything that we would from a client, meaning that we have a really powerful tool for prototyping and getting quick insights into our data.

Batch inserts using the shell

When using the shell, there will be many times we want to insert a large number of documents programmatically. The most straightforward implementation since we have a JavaScript shell is to iterate through a loop, generating each document along the way, and performing a write operation in every iteration in the loop, as follows:

```
> authorMongoFactory = function() {for(loop=0;loop<1000;loop++)
{db.books.insert({name: "MongoDB factory book" + loop})}}
function () {for(loop=0;loop<1000;loop++) {db.books.
insert({name: "MongoDB factory book" + loop})}}
```

In this simple example, we create an `authorMongoFactory()` method for an author who writes `1000` books on MongoDB with a slightly different name for each one:

```
> authorMongoFactory()
```

This will result in `1000` writes being issued to the database. While it is simple from a development point of view, this method will put a strain on the database.

Instead, using a `bulk` write, we can issue a single database `insert` command with the `1000` documents that we have prepared beforehand, as follows:

```
> fastAuthorMongoFactory = function() {
var bulk = db.books.initializeUnorderedBulkOp();
for(loop=0;loop<1000;loop++) {bulk.insert({name: "MongoDB
factory book" + loop})}}
bulk.execute();
}
```

The end result is the same as before, with the `1000` documents being inserted with the following structure in our `books` collection:

```
> db.books.find()
{ "_id" : ObjectId("59204251141daf984112d851"), "name" :
"MongoDB factory book0" }
{ "_id" : ObjectId("59204251141daf984112d852"), "name" :
"MongoDB factory book1" }
{ "_id" : ObjectId("59204251141daf984112d853"), "name" :
"MongoDB factory book2" }
...
{ "_id" : ObjectId("59204251141daf984112d853"), "name" :
"MongoDB factory book999" }
```

The difference from the user's perspective lies in the speed of execution and reduced strain on the database.

In the preceding example, we used `initializeUnorderedBulkOp()` for the `bulk` operation builder setup. The reason we did this is that we don't care about the order of insertions being the same as the order in which we add them to our `bulk` variable with the `bulk.insert()` command.

This makes sense when we can make sure that all operations are unrelated to each other or idempotent.

If we care about having the same order of insertions, we can use `initializeOrderedBulkOp()`; by changing the second line of our function, we get the following code:

```
var bulk = db.books.initializeOrderedBulkOp();
```

In the next section, we will see how we can use batch operations using the MongoDB shell to improve operational performance.

Batch operations using the mongo shell

In the case of inserts, we can generally expect that the order of operations doesn't matter.

The `bulk` command, however, can be used with many more operations than just inserts. In the following example, we have a single book with `isbn : 101` and a `name` value of `Mastering MongoDB` in a `bookOrders` collection with the number of available copies to purchase in the `available` field, with the `99` books available for purchase:

```
> db.bookOrders.find()
{ "_id" : ObjectId("59204793141daf984112dc3c"), "isbn" : 101,
"name" : "Mastering MongoDB", "available" : 99 }
```

With the following series of operations in a single `bulk` operation, we are adding one book to the inventory and then ordering `100` books, for a final total of 0 copies available:

```
> var bulk = db.bookOrders.initializeOrderedBulkOp();
> bulk.find({isbn: 101}).updateOne({$inc: {available : 1}});
> bulk.find({isbn: 101}).updateOne({$inc: {available : -100}});
> bulk.execute();
```

With the preceding code, we will get the following output:

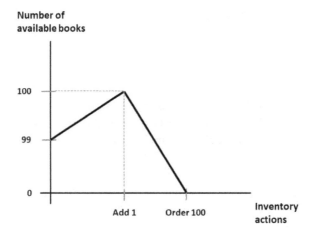

Figure 3.1 – Number of books over time

Using `initializeOrderedBulkOp()`, we can make sure that we are adding one book before ordering `100` so that we are never out of stock. On the contrary, if we were using `initialize-UnorderedBulkOp()`, we wouldn't have such a guarantee and we might end up with the `100`-book order coming in before the addition of the new book, resulting in an application error as we don't have that many books to fulfill the order.

When executing through an ordered list of operations, MongoDB will split the operations into batches of `1000` and group these by operation. For example, if we have `1004` inserts, `998` updates, `1004` deletes, and—finally—`5` inserts, we will end up with the following:

```
[1000 inserts]
[4 inserts]
[998 updates]
[1000 deletes]
[4 deletes]
[5 inserts]
```

The previous code can be explained as follows:

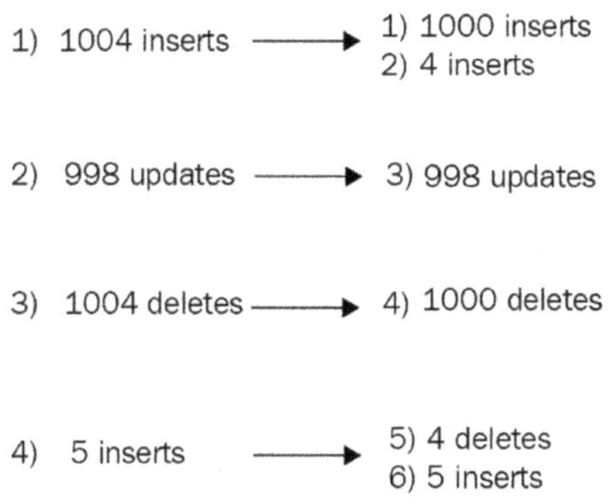

1) 1004 inserts ⟶ 1) 1000 inserts
                   2) 4 inserts

2) 998 updates ⟶ 3) 998 updates

3) 1004 deletes ⟶ 4) 1000 deletes

4) 5 inserts ⟶ 5) 4 deletes
               6) 5 inserts

Figure 3.2 – Bulk operations ordering

This doesn't affect the series of operations, but it implicitly means that our operations will leave the database in batches of `1000`. This behavior is not guaranteed to stay in future versions.

If we want to inspect the execution of a `bulk.execute()` command, we can issue `bulk.getOperations()` right after we type `execute()`.

> **Note**
>
> Since version 3.2, MongoDB has offered an alternative command for bulk writes: `bulkWrite()`.

`bulkWrite` arguments, as shown in the following code snippet, are the series of operations we want to execute; `WriteConcern` (the default is again `1`), and if the series of write operations should get applied in the order that they appear in the array (they will be ordered by default):

```
> db.collection.bulkWrite(
  [ <operation 1>, <operation 2>, ... ],
```

```
{
writeConcern : <document>,
ordered : <boolean>
}
)
```

The following operations are the same ones supported by `bulk`:

- `insertOne`

- `updateOne`

- `updateMany`

- `deleteOne`

- `deleteMany`

- `replaceOne`

`updateOne`, `deleteOne`, and `replaceOne` have matching filters; if they match more than one document, they will only operate on the first one. It's important to design these queries so that they don't match more than one document; otherwise, the behavior will be undefined.

In this section, we learned how we can script using the MongoDB shell. In the next section, we will go through administration tasks using the MongoDB shell.

Administration

Using MongoDB should, for the most part, be as transparent as possible to the developer. Since there are no schemas, there is no explicit need for migrations, and generally, developers find themselves spending less time on administrative tasks in the database world.

That said, there are several tasks that an experienced MongoDB developer or architect can perform to keep up the speed and performance of MongoDB.

Administration is generally performed on three different levels, ranging from more generic to more specific: **process**, **collection**, and **index**.

At the process level, there is the `shutDown` command to shut down the MongoDB server.

At the database level, we have the following commands:

- `dropDatabase` to drop the entire database

- `listCollections` to retrieve the collection names in the current database

- `copyDB` or `clone` to clone a remote database locally
- `repairDatabase` for when our database is not in a consistent state due to an unclean shutdown

In comparison, at the collection level, the following commands are used:

- `drop`: To drop a collection
- `create`: To create a collection
- `renameCollection`: To rename a collection
- `cloneCollection`: To clone a remote collection to our local database
- `cloneCollectionAsCapped`: To clone a collection into a new capped collection
- `convertToCapped`: To convert a collection to a capped one

At the index level, we can use the following commands:

- `createIndexes`: To create new indexes in the current collection
- `listIndexes`: To list existing indexes in the current collection
- `dropIndexes`: To drop all indexes from the current collection
- `reIndex`: To drop and recreate an index in the current collection

In the following sections, we will also go through a few other commands that are more important from an administration standpoint.

compact

MongoDB documents take up a specified amount of space on a disk. If we perform an update that increases the size of a document, this may end up being moved out of sequence to the end of the storage block, creating a hole in storage, resulting in increased execution times for this update, and possibly missing it from running queries. The `compact` operation will defragment space and result in less space being used.

We can update a document by adding an extra 10 bytes, showing how it will be moved to the end of the storage block, and creating an empty space in the physical storage, as illustrated in the following diagram:

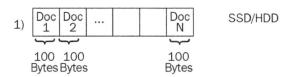

2) DOC 2 size=90 Bytes of data,10B padding

3) Update Doc 2. New size 110 Bytes

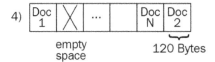

Figure 3.3 – How fragmentation occurs

compact can also take a force argument, as follows:

```
> db.runCommand ( { compact: '<collection>', force: true } )
```

The force parameter will force run the compact operation on the primary server of a replica set.

We can similarly add a parameter for a comment, like so:

```
{ comment: 'any description of the operation that we would like
to appear in the logs' }
```

currentOp() and killOp()

db.currentOp() will show us the currently running operation in the database and will attempt to kill it. We need to run the use admin command before running killOp(). Needless to say, using killOp() against internal MongoDB operations is not recommended or advised, as the database may end up in an undefined state. The killOp() command can be used like so:

```
> db.runCommand( { "killOp": 1, "op": <operationId> } )
```

collMod

collMod is used to pass flags to a collection by modifying the underlying database's behavior.

Since version 3.2, the most interesting set of flags that we can pass to a collection is document validation.

Document validation can specify a set of rules to be applied to new updates and inserts into a collection. This means that current documents will be checked if they get modified.

We can only apply validations to documents that are already valid if we set validationLevel to moderate. By specifying validationAction, we can log documents that are invalid by setting it to warn or prevent updates from happening altogether by setting it to error.

For example, with the previous example of bookOrders, we can set validator on the isbn and name fields being present for every insert or update, as demonstrated in the following code snippet:

```
> db.runCommand( { collMod: "bookOrders",
"validator" : {
          "$and" : [
                {
                    "isbn" : {
                        "$exists" : true
                    }
                },
                {
                    "name" : {
                        "$exists" : true
                    }
                }
          ]
      }
})
```

Here, we get back the following code:

```
{ "ok" : 1 }
```

Then, if we try to insert a new document with only the isbn field being present, we get an error, as shown here:

```
> db.bookOrders.insert({isbn: 102})
WriteResult({
"nInserted" : 0,
"writeError" : {
"code" : 121,
"errmsg" : "Document failed validation"
```

```
}
})
>
```

We get an error because our validation has failed. Managing validation from the shell is really useful as we can write scripts to manage it, and also make sure that everything is in place.

After learning how to administer our MongoDB server, in the next section, we will learn about the MapReduce framework.

MapReduce in the mongo shell

MongoDB allows users to write MapReduce scripts straight from the shell. This capability has not been supported as needed throughout the years and, as such, has now fallen behind the much more evolved aggregation framework, which we generally recommend over MapReduce scripting.

> **Important note**
> MapReduce has been deprecated in version 5 and should not be used for new development. It is recommended that we rewrite existing MapReduce code using the aggregation framework, which we will visit in *Chapter 7, Aggregation*.

MapReduce is a data processing method for getting aggregation results from large sets of data. The main advantage of this is that it is inherently parallelizable, as evidenced by frameworks such as Hadoop.

MapReduce is really useful when used to implement a data pipeline. Multiple MapReduce commands can be chained to produce different results. An example of this is aggregating data by using different reporting periods (such as hour, day, week, month, and year) where we use the output of each more granular reporting period to produce a less granular report.

A simple example of MapReduce being given a collection named `books` with the following data would look like this:

```
> db.books.find()
{ "_id" : ObjectId("592149c4aabac953a3a1e31e"), "isbn" : "101",
"name" : "Mastering MongoDB", "price" : 30 }
{ "_id" : ObjectId("59214bc1aabac954263b24e0"), "isbn" : "102",
"name" : "MongoDB in 7 years", "price" : 50 }
{ "_id" : ObjectId("59214bc1aabac954263b24e1"), "isbn" : "103",
"name" : "MongoDB for experts", "price" : 40 }
```

Our map and reduce functions are defined like this:

```
> var mapper = function() {
                    emit(this.id, 1);
            };
```

In this `mapper` variable, we simply output a key of `id` of each document with a value of `1`, as follows:

```
> var reducer = function(id, count) {
                    return Array.sum(count);
            };
```

In `reducer`, we sum across all values (where each one has a value of `1`), as follows:

```
> db.books.mapReduce(mapper, reducer, { out:"books_count" });
{
"result" : "books_count",
"timeMillis" : 16613,
"counts" : {
"input" : 3,
"emit" : 3,
"reduce" : 1,
"output" : 1
},
"ok" : 1
}
> db.books_count.find()
{ "_id" : null, "value" : 3 }
>
```

Our final output will be a document with no **identifier** (**ID**) since we didn't output any value for ID, and a value of 6, since there are six documents in the input dataset.

Using MapReduce, MongoDB will apply a map to each input document, emitting key-value pairs at the end of the map phase. Then, each reducer will get key-value pairs with the same key as the input, processing all multiple values. The reducer's output will be a single key-value pair for each key.

Optionally, we can use a finalize function to further process the results of mapper and reducer. MapReduce functions use JavaScript and run within the mongod process. MapReduce can output inline as a single document, subject to the 16 **megabyte** (**MB**) document size limit, or as multiple documents in an output collection. Input and output collections can be sharded.

MapReduce concurrency

MapReduce operations will place several short-lived locks that should not affect operations. However, at the end of the reduce phase, if we output the data to an existing collection, then output actions such as merge, reduce, and replace will take an exclusive global write lock for the whole server, blocking all other writes to the db instance. If we want to avoid this, then we should invoke mapReduce in the following way:

```
> db.collection.mapReduce(
                    mapper,
                    reducer,
                    {
                        out: { merge/reduce: bookOrders,
    nonAtomic: true    }
                    })
```

We can apply nonAtomic only to merge or reduce actions. replace will just replace the contents of documents in bookOrders, which will not take much time anyway.

With the merge action, the new result is merged with the existing result if an output collection already exists. If an existing document has the same key as the new result, then it will overwrite the existing document.

With the reduce action, the new result is processed together with the existing result if an output collection already exists. If an existing document has the same key as the new result, it will apply the reducer function to both the new and existing documents and will overwrite the existing document with the result.

Although MapReduce has been present since the early versions of MongoDB, it hasn't evolved as much as the rest of the database, resulting in its usage being less than that of specialized MapReduce frameworks such as Hadoop, which we will learn more about in *Chapter 12, Harnessing Big Data with MongoDB*.

Incremental MapReduce

Incremental MapReduce is a pattern where we use MapReduce to aggregate previously calculated values. An example of this could be counting non-distinct users in a collection for different reporting periods (that is, by hour, day, or month) without the need to recalculate the result every hour.

To set up our data for incremental MapReduce, we need to do the following:

- Output our reduce data to a different collection

- At the end of every hour, query only for the data that got into the collection in the last hour

- With the output of our reduce data, merge our results with the calculated results from the previous hour

Continuing with the previous example, let's assume that we have a `published` field in each of the documents with our input dataset, as shown in the following code snippet:

```
> db.books.find()
{ "_id" : ObjectId("592149c4aabac953a3a1e31e"), "isbn" : "101",
"name" : "Mastering MongoDB", "price" : 30, "published" :
ISODate("2017-06-25T00:00:00Z") }
{ "_id" : ObjectId("59214bc1aabac954263b24e0"), "isbn" : "102",
"name" : "MongoDB in 7 years", "price" : 50, "published" :
ISODate("2017-06-26T00:00:00Z") }
```

Using our previous example of counting books, we will get the following code:

```
var mapper = function() {
                        emit(this.id, 1);
                };
var reducer = function(id, count) {
                        return Array.sum(count);
                };
> db.books.mapReduce(mapper, reducer, { out: "books_count" })
{
"result" : "books_count",
"timeMillis" : 16700,
"counts" : {
"input" : 2,
"emit" : 2,
"reduce" : 1,
"output" : 1
```

```
    },
    "ok" : 1
    }
    > db.books_count.find()
    { "_id" : null, "value" : 2 }
```

Now, we get a third book in our `mongo_book` collection with a document, as follows:

```
    { "_id" : ObjectId("59214bc1aabac954263b24e1"), "isbn" : "103",
    "name" : "MongoDB for experts", "price" : 40, "published" :
    ISODate("2017-07-01T00:00:00Z") }
    > db.books.mapReduce( mapper, reducer, { query: { published:
    { $gte: ISODate('2017-07-01 00:00:00') } }, out: { reduce:
    "books_count" } } )
    > db.books_count.find()
    { "_id" : null, "value" : 3 }
```

What happened in the preceding code is that by querying for documents in July 2017, we only got the new document out of the query and then used its value to reduce the value with the already calculated value of 2 in our `books_count` document, adding 1 to the final sum of 3 documents.

This example, as contrived as it is, shows a powerful attribute of MapReduce: the ability to re-reduce results to incrementally calculate aggregations over time.

Troubleshooting MapReduce

Over the years, one of the major shortcomings of MapReduce frameworks has been the inherent difficulty in troubleshooting, as opposed to simpler non-distributed patterns. Most of the time, the most effective tool is debugging using `log` statements to verify that output values match our expected values. In the mongo shell, which is a JavaScript shell, it is as simple as providing the output using the `console.log()` function.

Diving deeper into MapReduce in MongoDB, we can debug both in the map and the reduce phase by overloading the output values.

By debugging the `mapper` phase, we can overload the `emit()` function to test what the output key values will be, as follows:

```
    > var emit = function(key, value) {
        print("debugging mapper's emit");
        print("key: " + key + "  value: " + tojson(value));
    }
```

We can then call it manually on a single document to verify that we get back the key-value pair that we expect, as follows:

```
> var myDoc = db.orders.findOne ( { _id:
ObjectId("50a8240b927d5d8b5891743c") } );
> mapper.apply(myDoc);
```

The `reducer` function is somewhat more complicated. A MapReduce `reducer` function must meet the following criteria:

- It must be idempotent

- It must be commutative

- The order of values coming from the `mapper` function should not matter for the reducer's result

- The `reducer` function must return the same type of result as the `mapper` function

We will dissect each of these following requirements to understand what they really mean:

- **It must be idempotent**: MapReduce, by design, may call the `reducer` function multiple times for the same key with multiple values from the `mapper` phase. It also doesn't need to reduce single instances of a key as it's just added to the set. The final value should be the same no matter the order of execution. This can be verified by writing our own `verifier` function and forcing `reducer` to re-reduce, or by executing `reducer` many times, as shown in the following code snippet:

  ```
  reduce( key, [ reduce(key, valuesArray) ] ) == reduce(
  key, valuesArray )
  ```

- **It must be commutative**: As multiple invocations of the `reducer` function may happen for the same `key` value, if it has multiple values, the following code should hold:

  ```
  reduce(key, [ C, reduce(key, [ A, B ]) ] ) == reduce(
  key, [ C, A, B ] )
  ```

- **The order of values coming from the mapper function should not matter for the reducer's result**: We can test that the order of values from `mapper` doesn't change the output for `reducer` by passing in documents to `mapper` in a different order and verifying that we get the same results out, as follows:

  ```
  reduce( key, [ A, B ] ) == reduce( key, [ B, A ] )
  ```

- **The reducer function must return the same type of result as the mapper function**: Hand-in-hand with the first requirement, the type of object that the `reducer` function returns should be the same as the output of the `mapper` function.

The MapReduce framework is deprecated in version 5 but there is still legacy code that needs to be understood and rewritten. In the next section, we will learn about the recommended way to rewrite it: the aggregation framework.

Aggregation framework

Since version 2.2, MongoDB has provided a better way to work with aggregation—one that has been supported, adopted, and enhanced regularly ever since. The aggregation framework is modeled after data processing pipelines.

In data processing pipelines, there are three main operations: filters that operate like queries, filtering documents, and document transformations that transform documents to get them ready for the next stage. In the following sections, we will discuss how these three methods map to each other on a high level.

SQL to aggregation

An aggregation pipeline can replace and augment querying operations in the shell. A common pattern for development goes like this:

- Verifying that we have the correct data structures and getting quick results using a series of queries in the shell
- Prototyping pipeline results using the aggregation framework
- Refining and refactoring if/when needed, either by **extract, transform, load** (ETL) processes to get data into a dedicated data warehouse or by more extensive usage of the application layer to get the insights that we need

In the following table, we can see how **Structured Query Language** (SQL) commands map to aggregation framework operators:

SQL	Aggregation framework
WHERE/HAVING	$match
GROUP BY	$group
SELECT	$project
ORDER BY	$sort
LIMIT	$limit
sum()/count()	$sum
Join	$lookup

Table 3.2 – SQL to aggregation framework mapping

Aggregation versus MapReduce

In MongoDB, we can essentially get data out of our database by using three methods: querying, the aggregation framework, and MapReduce. All three of them can be chained to each other, and many times it is useful to do so; however, it's important to understand when we should use aggregation and when MapReduce may be a better alternative.

> **Note**
> We can use both aggregation and MapReduce with sharded databases.

Aggregation is based on the concept of a pipeline. As such, it's important to be able to model our data from the input to the final output, in a series of transformations and processing that can get us there. It's also mostly useful when our intermediate results can be used on their own, or feed parallel pipelines. Our operations are limited by the operators that we have available from MongoDB, so it's important to make sure that we can calculate all the results we need by using the available commands.

MapReduce, on the other hand, can be used to construct pipelines by chaining the output of one MapReduce job to the input of the next one via an intermediate collection, but this is not its primary purpose.

MapReduce's most common use case is to periodically calculate aggregations for large datasets. Having MongoDB's querying in place, we can incrementally calculate these aggregations without the need to scan through the whole input table every time. In addition, its power comes from its flexibility, as we can define mappers and reducers in JavaScript with the full flexibility of the language when calculating intermediate results. Not having the operators that the aggregation framework provides us with, we have to implement them on our own.

In many cases, the answer is not either/or. We can (and should) use the aggregation framework to construct our ETL pipeline and resort to MapReduce for the parts that are not yet supported sufficiently by it.

A complete use case with aggregation and MapReduce is provided in *Chapter 7, Aggregation*.

Securing the shell

MongoDB is a database developed with ease of development in mind. As such, security at the database level was not baked in from the beginning, and it was up to the developers and administrators to secure the MongoDB host from accessing outside the application server.

Unfortunately, this means that as far back as 2015, there were 39,890 databases found open to the internet with no security access configured. Many of them were production databases, one belonging to a French telecom operator and containing more than 8 million records from its customers.

Nowadays, there is no excuse for leaving any MongoDB server with the default authentication off settings at any stage of development, from local server deployment to production.

Authentication and authorization

Authentication and authorization are closely connected and sometimes confused. Authentication is about verifying the identity of a user to the database. An example of authentication is **Secure Sockets Layer** (**SSL**), where the web server verifies its identity—that it is who it claims to be—to the user.

Authorization is about determining which actions a user can take on a resource. In the next sections, we will discuss authentication and authorization with these definitions in mind. We will also examine some security tips for MongoDB as of the most current version.

Authorization with MongoDB

MongoDB's most basic authorization relies on the username/password method. By default, MongoDB will not start with authorization enabled. To enable it, we need to start our server with the `--auth` parameter, as follows:

```
$ mongod --auth
```

To set up authorization, we need to start our server without authorization to set up a user. Setting up an `admin` user is simple, as we can see here:

```
> use admin
> db.createUser(
  {
    user: <adminUser>,
    pwd: <password>,
    roles: [ { role: <adminRole>, db: "admin" } ]
  }
)
```

Here, `<adminUser>` is the name of the user we want to create, `<password>` is the password, and `<adminRole>` can be any of the following values ordered from the most powerful to the least powerful, as shown in the following list:

- `root`
- `dbAdminAnyDatabase`
- `userAdminAnyDatabase`
- `readWriteAnyDatabase`
- `readAnyDatabase`
- `dbOwner`
- `dbAdmin`

- userAdmin

- readWrite

- read

Of these roles, root is the superuser that allows access to everything. This is not recommended to be used, except for special circumstances.

All the AnyDatabase roles provide access to all databases, of which dbAdminAnyDatabase combines the userAdminAnyDatabase and readWriteAnyDatabase scopes being an adminstrator again in all databases.

The rest of the roles are defined in the database that we want them to apply, by changing the roles subdocument of the preceding db.createUser(); for example, to create a dbAdmin role for our mongo_book database, we would use the following code:

```
> db.createUser(
{
user: <adminUser>,
pwd: <password>,
roles: [ { role: "dbAdmin", db: "mongo_book" } ]
}
)
```

Cluster administration has even more roles, which we will cover in more depth in *Chapter 13, Mastering Replication.*

Finally, when we restart our database with the --auth flag set, we can use either the command line or the connection string (from any driver) to connect as admin and create new users with predefined or custom-defined roles, as follows:

```
mongodb://[username:password@]host1[:port1][,host2[:port2],...
[,hostN[:portN]]][/[database][?options]]
```

Authentication and authorization are essential prerequisites for any production-level MongoDB system. In the next section, we will establish a baseline of security best practices.

Security tips for MongoDB

Common software system security precautions apply with MongoDB. We will outline some of them in the following sections and learn how to enable them.

Encrypting communication using Transport Layer Security (TLS)/SSL

Communication between the mongod or mongos server and the client mongo shell or applications should be encrypted. This is supported in most MongoDB distributions from version 3.0 onward; however, we need to take care that we download the proper version with SSL support.

After this, we need to get a signed certificate from a trusted **certificate authority** (**CA**) or sign our own. Using self-signed certificates is fine for preproduction systems, but in production it will mean that MongoDB servers won't be able to verify our identity, leaving us susceptible to **man-in-the-middle** (**MitM**) attacks; thus, using a proper certificate is highly recommended.

To start our MongoDB server with SSL, we need the following code:

```
$ mongod --sslMode requireSSL --sslPEMKeyFile <pem> --sslCAFile
<ca>
```

Here, `<pem>` is our `.pem` signed certificate file, and `<ca>` is the `.pem` root certificate from the CA that contains the root certificate chain.

These options can also be defined in our `mongod.conf` or `mongos.conf` configuration file in **YAML Ain't Markup Language** (**YAML**) file format, as follows:

```
net:
  ssl:
      mode: requireSSL
      PEMKeyFile: /etc/ssl/mongodb.pem
      CAFile: /etc/ssl/ca.pem
      disabledProtocols: TLS1_0,TLS1_1,TLS1_2
```

Here, we specified a `PEMKeyFile` instance, a `CAFile` instance, and also that we won't allow the server to start with certificates that follow the `TLS1_0`, `TLS1_1`, or `TLS1_2` versions. These are the available versions for `disabledProtocols` at this time.

Encrypting data

Using WiredTiger is highly recommended for encrypting data at rest, as it supports it natively from version 3.2.

For users of the Community Edition version, this can be achieved in the storage selection of their choice; for example, in **Amazon Web Services** (**AWS**) using **Elastic Block Store** (**EBS**) encrypted storage volumes.

> **Note**
> This feature is available only for MongoDB Enterprise Edition.

Limiting network exposure

The oldest security method to secure any server is to disallow it from accepting connections from unknown sources. In MongoDB, this is done in a configuration file with a simple line, as follows:

```
net:
  bindIp: <string>
```

Here, `<string>` is a comma-separated list of **Internet Protocol** (**IP**) addresses that the MongoDB server will accept connections from.

Firewalls and VPNs

Together with limiting network exposure on the server side, we can use firewalls to prevent access to our network from the outside internet. **Virtual private networks** (**VPNs**) can also provide tunneled traffic between our servers, but regardless, they shouldn't be used as our sole security mechanism.

Auditing

No matter how secure any system is, we need to keep a close eye on the system from an auditing perspective to make sure that we detect possible breaches and stop them as soon as possible.

> **Note**
> This feature is available only for MongoDB Enterprise Edition.

For users of the Community Edition version, we have to set up auditing manually by logging changes to documents and collections in the application layer, possibly in a different database altogether. This will be addressed in the next chapter, which covers advanced querying using client drivers.

Using secure configuration options

It goes without saying that the same configuration options should be used. We must use one of the following:

- MapReduce
- The mongo shell group operation or a group operation from our client driver
- `$where` JavaScript server evaluation

We can use the `-noscripting` command-line parameter on startup to disable server-side scripting, and it's a good practice to use it to prevent accidental server-side evaluation when possible. The mongo shell group operation, as mentioned in the previous list, can be a tricky one as many drivers may use MongoDB's `group()` command when we issue group commands in the driver. However, given the limitations that `group()` has in terms of performance and output documents, we should rethink our design to use the aggregation framework or application-side aggregations.

The web interface also has to be disabled by not using any of the following commands:

- `net.http.enabled`

- `net.http.JSONPEnabled`

- `net.http.RESTInterfaceEnabled`

On the contrary, `wireObjectCheck` needs to remain enabled as it is by default. `WireObjectCheck` guarantees that every document stored is valid **Binary JSON (BSON)**. After learning how to secure the MongoDB shell, we will learn next about the different authentication methods supported by MongoDB.

Authentication with MongoDB

By default, MongoDB uses **Salted Challenge Response Authentication Mechanism secure hashing algorithm 1 (SCRAM-SHA-1)** as the default challenge-and-response authentication mechanism. This is an SHA-1 username/password-based mechanism for authentication. All drivers and the mongo shell itself have built-in methods to support it. The Enterprise Edition supports more authentication options, which we will cover in the next sections.

> **Note**
> The authentication protocol in MongoDB has changed since version 3.0. In older versions, the less secure MONGODB-CR was used.

Enterprise Edition

MongoDB's Enterprise Edition is a paid subscription product offering more features around security and administration. The two most famous ones are Kerberos and **Lightweight Directory Access Protocol (LDAP)** authentication, which we will visit in the next sections.

Kerberos authentication

MongoDB Enterprise Edition also offers Kerberos authentication. Kerberos is named after the three-headed guard dog of the God of the underworld, Hades, in Greek mythology. Kerberos provides mutual authentication between client and server, protecting against eavesdropping and replay attacks.

Kerberos is widely used in Windows systems through integration with Microsoft's **Active Directory (AD)**. To install Kerberos, we need to start `mongod` without Kerberos set up, then connect to the `$external` database (not the adminstrator that we normally use for adminstrator authorization), and create a user with a Kerberos role and permissions, as follows:

```
use $external
db.createUser(
    {
```

```
      user: "mongo_book_user@packt.net",
      roles: [ { role: "read", db: "mongo_book" } ]
   }
)
```

In the preceding example, we authorize the mongo_book_user@packt.net user to read our mongo_book database, just as we would do with a user using our admin system.

After that, we need to start our server with Kerberos support by passing in the authenticationMechanisms parameter, as follows:

```
--setParameter authenticationMechanisms=GSSAPI
```

Now, we can connect from our server or command line, as follows:

```
$ mongo.exe --host <mongoserver>
--authenticationMechanism=GSSAPI
--authenticationDatabase='$external' --username mongo_book_
user@packt.net
```

LDAP authentication

Similar to Kerberos authentication, we can also use LDAP in MongoDB Enterprise Edition only. The user setup needs to be done in the $external database and must match the name of the authentication LDAP name. The name may need to pass through a transformation, and this may cause a mismatch between the LDAP name and the user entry in the $external database.

Setting up LDAP authentication is beyond the scope of this book, but the important thing to consider is that any changes in the LDAP server may need tweaks in the MongoDB server, which won't happen automatically.

The new mongosh shell

Starting from version 5, MongoDB introduced a new shell, **mongosh**, which we have been using throughout the chapter.

mongosh is the recommended shell going forward. mongosh is backward compatible with the older mongo shell.

Every command of the mongo shell that is implemented in the mongosh shell will have the exact same syntax, but not all commands are implemented yet.

The main advantages of the new shell are set out here:

- Improved syntax highlighting

- Improved command history

- Improved logging

mongosh is a fully functional JavaScript and Node.js 14.x **Read-Eval-Print-Loop** (**REPL**) environment that can be used for query and operations testing.

MongoDB Stable API

Starting from version 5, MongoDB introduced the Stable API. The Stable API provides a guarantee that the API will not break for client-server communication. The Stable API is declared when using any driver or the mongosh shell, in a similar fashion to the following mongosh example:

```
--apiVersion 1
```

> **Note**
> 1 is the only API version available as of MongoDB 6.0.

StableAPI guarantees backward compatibility between MongoDB server upgrades.

This means that we can continue upgrading our MongoDB server without any significant risk that our application connected to the MongoDB server will behave differently.

This guarantee holds correct under the following three constraints:

- We need to declare apiVersion in the client

- We need to use a supported version of the official MongoDB client

- We can only use commands and features that are supported in this API version

Following the third constraint, as of apiVersion=1, we can only use any of the following commands:

- abortTransaction: To terminate a multi-document (also known as distributed) transaction and roll back its results.

- aggregate: (*with limitations*) To execute an aggregation pipeline.

- authenticate: To authenticate a client using the x.509 authentication mechanism.

- count: Introduced in version 6 and available since version 5.0.9, this counts the number of documents in a collection or a view.

- collMod: To modify view definitions or add options to a collection.

- `commitTransaction`: To commit a multi-document transaction.
- `create`: (*with limitations*) To create a collection or view.
- `createIndexes`: (*with limitations*) To create one or more indexes in a collection.
- `delete`: To remove one or more documents from a collection.
- `drop`: To remove an entire collection.
- `dropDatabase`: To remove an entire database.
- `dropIndexes`: To remove one or more indexes from a collection.
- `endSessions`: To expire specific sessions after waiting for the timeout period.
- `explain`: (*output may change in future versions*) To get an execution query plan for MongoDB operations.
- `find`: (*with limitations*) To execute a query against a collection.
- `findAndModify`: To execute a query against a collection and modify one or more documents in the result set.
- `getMore`: To fetch more documents in commands that use a cursor to return results in batches.
- `insert`: To insert one or more documents in a collection.
- `hello`: To return information about the MongoDB server. This may include primary/secondary replica set information as well as authentication method supported and other role-level information.
- `killCursors`: To delete cursors that are returned from queries that return results in batches.
- `listCollections`: To list collections in the database.
- `listDatabases`: To list databases.
- `listIndexes`: To list indexes in a collection.
- `ping`: To ping a server, equivalent to the **Internet Control Message Protocol** (**ICMP**) echo request/reply.
- `refreshSessions`: To update the last used time for specified sessions in order to extend their active state.
- `update`: To update one or more documents in a collection.

The methods mentioned here that have the note (*with limitations*) are only partially supported by the Stable API guarantee. This should be taken into account when overlaying our code on top of the Stable API. We can also set the `apiStrict` Boolean flag to `True` to prohibit clients from using commands that are not whitelisted. MongoDB will return an `apiStrictError` in such cases.

> **Note**
>
> `apiStrict` defaults to `False`.

Finally, we can pass the `apiDeprecationErrors` Boolean parameter value to `True` in order to invoke an error when trying to access any functionality that is deprecated.

Summary

In this chapter, we scratched the tip of the iceberg of CRUD operations. Starting from the mongo shell, we learned how to insert, delete, read, and modify documents. We also discussed the differences between one-off inserts and inserting in batches for performance.

Following that, we discussed administration tasks and how to perform them in the mongo shell. MapReduce and its successor, aggregation framework, were also discussed in this chapter, including how they compare, how to use them, and how we can translate SQL queries to aggregation framework pipeline commands.

Finally, we discussed security and authentication with MongoDB, the new versioning scheme, and the new shell, mongosh. Securing our database is of paramount importance; we will learn more about this in *Chapter 9, Monitoring, Backup, and Security*.

In the next chapter, we will learn more about auditing and MongoDB.

4
Auditing

In this chapter, we will learn how to set up, configure, and deploy auditing using MongoDB Enterprise Edition to showcase on-premises deployment and MongoDB Atlas for cloud-based deployment. Auditing is a mandatory requirement in regulated industries and is part of any best practice guideline around best security practices. By the end of this chapter, we will know how to deploy auditing on-premises or in a MongoDB Atlas environment, enabling us to design secure systems and comply with regulatory requirements if needed.

Auditing is a MongoDB feature that is only available in MongoDB Enterprise Edition and, at the time of writing, the non-shared instances of MongoDB Atlas, the M10 size or larger.

In this chapter, we will cover the following topics:

- Auditing and logging differences
- Audit setup in MongoDB Enterprise Edition
- Audit setup in MongoDB Atlas, the cloud-based DBaaS offering by MongoDB
- Audit case study

Technical requirements

To follow along with the code in this chapter, you will need to install MongoDB Enterprise Edition locally or connect to a MongoDB Atlas database. You can download MongoDB Enterprise Edition for developer evaluation for free from mongodb.com or use the fully managed DBaaS MongoDB Atlas offering, which provides a free tier and seamless upgrades to the latest version.

Auditing and logging differences

Auditing and **logging** are two concepts that are, sometimes, used interchangeably in some contexts. However, there are some important differences between the two.

Logging refers to program-level events that happen throughout its execution. Usually, these events are atomic operations that happen internally and contain information that is useful for development, bug identification, and fixing. Logging includes information about what happened in the level of detail that is useful for the developers. Often, logs are deleted after a short- or developer-determined amount of time.

As an example of logging, when we start up the MongoDB server, we can see entries in the logs like the following:

```
{"t":{"$date":"2022-02-
05T13:10:05.951+00:00"},"s":"I",  "c":"STOR-
AGE", "id":22430,   "ctx":"Checkpointer","msg":"Wired-
Tiger message","attr":{"message":"[1644066605:951809]
[12592:0x700005bf7000], WT_SESSION.checkpoint: [WT_VERB_CHECK-
POINT_PROGRESS] saving checkpoint snapshot min: 37, snapshot
max: 37 snapshot count: 0, oldest timestamp: (0, 0) , meta
checkpoint timestamp: (0, 0) base write gen: 1"}
```

This log entry, which is from the preceding `WiredTiger` storage subsystem, is referring to an internal checkpoint save that is used for journaling purposes. This log entry can also be useful for debugging purposes if there is ever a system failure, but it is not so useful for auditing purposes.

On the other hand, auditing refers to business-level events.

A business-level event refers to an action that is usually performed by a user and refers to a domain rather than an implementation-level library.

Auditing answers the question of *"Who did What and When?"* in the system. We might also want to answer the question of *"Why?"* in the case of a root cause investigation.

As an example of auditing, the following is a line from the audit log in MongoDB:

```
{ "atype" : "createCollection", "ts" : { "$date" : "2022-
02-12T14:12:30.060+00:00" }, "uuid" : { "$binary" :
"2jmqHHr3SoirJJ1/KwHk+Q==", "$type" : "04" }, "local"
: { "ip" : "XXX", "port" : 27017 }, "remote" : { "ip" :
"86.153.147.144", "port" : 60764 }, "users" : [ { "user"
: "admin", "db" : "admin" } ], "roles" : [ { "role" :
"readWriteAnyDatabase", "db" : "admin" } ], "param" : { "ns" :
"audit-project.contracts" }, "result" : 0 }
```

Auditing configuration is usually driven by legal requirements in the case of certain industries and applications such as banking and defense.

Auditing is not performed in a vacuum but aims to offer a holistic view of a user's interaction with the system. We should be able to answer the fundamental question of *"What was the series of events that brought the system into the target state?"* using auditing.

Auditing can be kept indefinitely and will be kept for as long as the legal requirements dictate. Usually, auditing requirements derive from regulatory requirements and, as such, are quite prescriptive in the exact details of implementation. What to audit, how long to store the data, and how exactly we should store the data and other parameters are prescribed in specifications such as the **Payment Card Industry Data Security Standard** (**PCI DSS**), which is used for payments processing.

Generally, we use logging to perform auditing tasks.

Currently, auditing is only available in the MongoDB Enterprise Edition and MongoDB Atlas standalone instances. Percona Server for MongoDB also supports auditing in a way that is mostly compatible with MongoDB's own implementation.

Additionally, Amazon's DocumentDB implements auditing features via Amazon CloudWatch. At the time of writing, DocumentDB is compatible with MongoDB 3.6 and 4.0.

Audit setup in MongoDB Enterprise Edition

MongoDB Enterprise Edition is the paid version of MongoDB. Anyone can download it from the MongoDB site for development or evaluation purposes, but it needs a license before installing on production servers.

> **Note**
>
> MongoDB Enterprise Edition supports audit log encryption using an external **Key Management Interoperability Protocol** (**KMIP**) server, starting from version 6.0.

All of the following examples can be reproduced locally using MongoDB Enterprise Edition.

Audit events and format

We can record any of the following events in the console, the syslog, a JSON file, or a BSON file:

- Schema (DDL) such as creating or dropping a collection
- Replica set and sharded cluster operations such as sharding an existing collection or adding a new shard
- Authentication and authorization such as authentication failure events
- CRUD operations, such as inserting, updating, deleting, or querying documents in the collection

The audit message format is as follows:

```
{
    atype: <string>,
```

```
    ts : { $date: <timestamp> },
    uuid : { $binary: <string>, $type: <string> },
    local: { ip: <string>, port: <int> || isSystemUser: <boolean>
|| unix: <string> },
    remote: { ip: <string>, port: <int> || isSystemUser:
<boolean> || unix: <string> },
    users : [ { user: <string>, db: <string> }, ... ],
    roles: [ { role: <string>, db: <string> }, ... ],
    param: <document>,
    result: <int>
}
```

By default, fields such as ts (timestamp), users, and rules are included in every audit log.

The atype, param, and result values for the events that are audited are specified in the *Audit Event Actions, Details, and Results* section of the official MongoDB documentation, which can be found at https://www.mongodb.com/docs/manual/reference/audit-message/.

An example of an audit output message is creating a database. Using the JSON output notation, the relevant audit log would be as follows:

```
{
atype: 'createDatabase',
param: { ns: <database> },
result: 0,
ts : { $date: <timestamp> },
uuid : { $binary: <string>, $type: <string> },
local: { ip: <string>, port: <int> || isSystemUser: <boolean>
|| unix: <string> },
remote: { ip: <string>, port: <int> || isSystemUser: <boolean>
|| unix: <string> },
users : [ { user: <string>, db: <string> }, ... ],
roles: [ { role: <string>, db: <string> }, ... ],
}
```

Here, the param field is used to add metadata and details of the audited event. In this case, the embedded subdocument contains the ns field, which has a string value with the database name.

In the `createCollection` audit event, we can audit the creation of a collection or a view. If we create a view, then the `param` field will contain the aggregation pipeline definition for the view, which is useful if we want to reproduce the view for testing or debugging purposes, as shown in the following code:

```
{
    ns: <database>.<collection || view>,
    viewOn: <database>.<collection>,
    pipeline: [ <pipeline definition> ]
}
```

Following on from audit events logging, in the next section, we will learn about audit filters.

Audit filters

By default, MongoDB will record all auditable operations, as detailed in the *Audit Event Actions, Details, and Results* section of the previously mentioned documentation, in the output stream as per our configuration settings.

We can use audit filters to limit the events that will be audited. The filters can either be applied statically on server startup or dynamically at runtime.

To start our server with audit filters, we can either define them via the command line or through the `mongod` or `mongos config` file.

Here is an example of how to set up static audit filters on startup, using the command line:

```
mongod --auditFilter '{ atype: "authenticate", "param.db":
"test" }'
```

Here is an example using a `mongod` or `mongos config` file:

```
auditLog:
  filter: '{ atype: "authenticate", "param.db": "test" }'
```

Audit filters are useful, even more so if we don't need to restart our server to edit them. We can enable runtime audit filter editing by starting our server with `auditLog.runtimeConfiguration=true` and unsetting (not declaring at all) both `auditLog.filter` and `auditAuthorizationSuccess` variables.

If either of these two variables is set, then the server will fail to start. The reason for this is that MongoDB cannot decide between using `runtimeConfiguration` or the configuration provided at startup.

The following examples show how we can use dynamically configured audit filters at runtime.

To edit the `auditAuthorizationSuccess` variable during runtime, we can issue an admin command directly from the mongosh shell, as follows:

```
db.adminCommand( { setParameter: 1, auditAuthorizationSuccess:
true } )
```

Similarly, we can edit the `auditlog` filter with another `db.adminCommand()` command. Here is an example:

```
db.adminCommand( { setAuditConfig: 1, filter: { atype:
"authenticate", "param.db": "test" } } )
```

This is the equivalent of the example used in the first part of this section, where we defined audit filters at startup time. These filters are commonly referred to as static audit filters.

Audit guarantee

MongoDB will always write the audit event even before writing to the journal for all operations that affect the database's durability, such as any document create, update, and delete actions.

Audit events are guaranteed to have the same ordering as their operations for every connection. MongoDB guarantees that every event will be written to disk only after all prior events have been written.

In the next section, we will move from the MongoDB local setup to the auditing setup for MongoDB Atlas, which is the DBaaS offering from MongoDB.

Audit setup in MongoDB Atlas

MongoDB Atlas supports auditing in all dedicated clusters (non-free) from M10 and up.

At the time of writing, serverless instances are still in preview and do not support database auditing.

Enabling auditing in MongoDB Atlas is as simple as enabling the toggle button under **Security | Advanced settings**.

The following screenshot shows the different users and roles that we can configure to target for auditing:

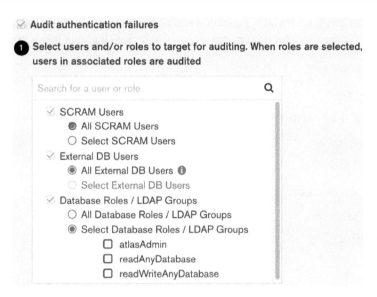

Figure 4.1 – Auditing users in MongoDB Atlas

Auditing can be configured on both a user level and a role level. Selecting a role will implicitly apply our selection to all users that are members of this role.

The list of roles is extensive but most of them are self-explanatory, such as `readWriteAnyDatabase`.

The following screenshot shows a snippet of the available actions that we can select to audit using MongoDB Atlas:

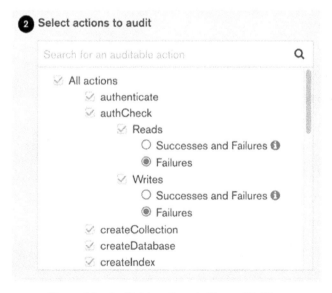

Figure 4.2 – Auditable actions in MongoDB Atlas

Additionally, we can configure auditing per action and audit a limited set of all database operations. The list of database operations that we can include or exclude from auditing logs is as follows:

- `authenticate`
- `authCheck`
- `createCollection`
- `createDatabase`
- `createIndex`
- `renameCollection`
- `dropCollection`
- `dropDatabase`
- `dropIndex`
- `createUser`
- `dropUser`
- `dropAllUsersFromDatabase`
- `updateUser`
- `grantRolesToUser`
- `revokeRolesFromUser`
- `createRole`
- `updateRole`
- `dropRole`
- `dropAllRolesFromDatabase`
- `grantRolesToRole`
- `revokeRolesFromRole`
- `grantPrivilegesToRole`
- `revokePrivilegesFromRole`
- `enableSharding`
- `shardCollection`
- `addShard`
- `removeShard`

- shutdown
- applicationMessage

Most of these preceding commands are self-explanatory.

A slight exception is `authcheck`. Locally, we need to set `auditAuthorizationSuccess` to true in order to log successful authorization attempts in the logs. This `Boolean` value works in conjunction with the `authcheck` event action.

In MongoDB Atlas, we get four radio buttons to choose from in `authcheck`. They are mapped to the following four actions:

- `authChecksReadFailures`
- `authChecksReadAll`
- `authChecksWriteFailures`
- `authChecksWriteAll`

Likewise, in a local deployment, `AuthChecksReadAll` and `authChecksWriteAll` will also set `auditAuthorizationSuccess` to true.

Setting `auditAuthorizationSuccess` to true applies to both reads and writes. There is no capability to audit reads and writes at a different granularity with regard to auditing successful events.

> **Important Note**
> Enabling auditing to log each and every successful login can severely impact performance.

JSON audit filter

The audit filter is the configuration file that specifies which events will be audited. The **Graphical User Interface (GUI)** driven filter builder is one way to configure MongoDB. The other way is through a JSON configuration file.

This is as simple as clicking on the relevant button:

USE CUSTOM JSON FILTER

Figure 4.3 – The custom audit filter button

Clicking on this button will allow us to view the JSON filter that results from our GUI changes. Then, we can edit the JSON file and toggle between the GUI and JSON representations of the filter.

Audit logging access

Audit events can be accessed via logs or a REST API.

A typical audit log line is shown as follows:

```
{ "atype" : "authCheck", "ts" : { "$date" : "2022-
02-10T07:17:44.892+00:00" }, "uuid" : { "$binary" :
"isBORpArSUCG4vONSn4Mqw==", "$type" : "04" }, "local" : {
"ip" : "192.168.253.134", "port" : 27017 }, "remote" : { "ip"
: "192.168.253.134", "port" : 34142 }, "users" : [], "roles"
: [], "param" : { "command" : "isMaster", "ns" : "admin",
"args" : { "isMaster" : 1, "topologyVersion" : { "processId"
: { "$oid" : "6204bc17fcb99d4cac30d5a0" }, "counter" : {
"$numberLong" : "3" } }, "maxAwaitTimeMS" : { "$numberLong"
: "10000" }, "$db" : "admin", "$readPreference" : { "mode" :
"primaryPreferred" } } }, "result" : 0 }
```

We can download the logs using the GUI from the project's dashboard page, as follows:

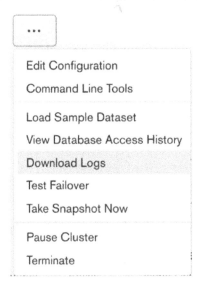

Figure 4.4 – The Download Logs option

Additionally, we can access the logs via the REST API.

To access the logs via the REST API, we need the following four parameters:

- The public API key
- The private API key

- The project ID

- The MongoDB cluster hostname

The public and private API keys are both generated from the **Organization | Access Manager page "create API" functionality** setting. The private key is only visible once, so we need to make a note of it.

Typically, the project ID is a long hex string, for example, 620173c921b1ab3de3e8e610, which we can retrieve from the **Organization | Projects** page.

To retrieve the project ID, click on the ellipsis in the project's row and select **Copy Project ID**, as shown in the following screenshot:

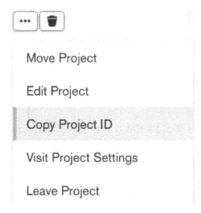

Figure 4.5 – Use the Copy Project ID option to retrieve the project ID

We are able to retrieve the project's ID.

Finally, the MongoDB cluster hostname is visible on the **Database** overview page.

Putting all the preceding steps together, the following `curl` command will download MongoDB's audit logs locally:

```
curl --user '{public_key:private_key}' --digest --header
'Accept: application/gzip' --request GET "https://cloud.
mongodb.com/api/atlas/v1.0/groups/{project_id}/clusters/
{hostname}/logs/{mongodb_log_filename}" --output "mongodb-
audit-log.gz"
```

The preceding `{mongodb_log_filename}` placeholder can have any of the following four values. These map to the four available log files in MongoDB Atlas:

- `mongodb.gz`

- `mongos.gz`

- `mongodb-audit-log.gz`
- `mongos-audit-log.gz`

In the previous sections, we learned about audit setup and configuration in MongoDB Enterprise Edition locally, along with how to use the cloud MongoDB Atlas offering. In the next section, we will put everything into action in a case study using MongoDB Atlas and the MongoDB **Command-Line Interface (CLI)**.

Audit case study

In this section, we will put everything together and present an audit project using MongoDB Atlas and the MongoDB CLI.

MongoDB CLI

The MongoDB CLI allows us to configure Atlas, Cloud Manager, Ops Manager, and IAM operations. In this section, we will use the MongoDB Atlas interface to connect to our Atlas instance.

We can install the MongoDB Atlas tool by following the instructions in the documentation page: `https://www.mongodb.com/docs/mongocli/stable/install/` or using the GitHub repo `https://github.com/mongodb/mongodb-atlas-cli`

We can use the general purpose mongocli or the Atlas specific atlascli.

The next step is to initiate and configure `mongocli` to authenticate to our cluster:

```
$ mongocli config
? Public API Key: [public_key]
? Private API Key: [? for help] ******************************
******
? Choose a default organization: [alex's Org]
? Choose a default project: [audit-project
(620173c921b1ab3de3e8e610)]
? Default MongoDB Shell Path: [/usr/local/bin]
```

You need to change the preceding input values marked with [] (for example, [public_key]) with your own values. Public and private API keys can be fetched from **Project** | **Access manager** in the MongoDB Atlas GUI.

The other inputs, namely `default organization`, `default project`, and `Default MongoDB Shell Path` should have the correct default values auto-discovered by MongoDB.

On completing this setup wizard, we will be authenticated into the cluster from the command line.

> **Important Note**
>
> By default, the MongoDB CLI will store the configuration settings in ~/.config/mongocli.toml.

Then, we can access our Atlas cluster from the shell, and we are able to script any interaction with it:

```
$ mongocli atlas dbusers list
[
  {
    "databaseName": "admin",
    "ldapAuthType": "NONE",
    "x509Type": "NONE",
    "awsIAMType": "NONE",
    "groupId": "620173c921b1ab3de3e8e610",
    "roles": [
      {
        "roleName": "readWriteAnyDatabase",
        "databaseName": "admin"
      }
    ],
    "scopes": [],
    "username": "admin"
  }
]
```

We should then enable access to our database from our IP address. By default, MongoDB Atlas databases are not open to external connections, including the admin IP. To change that, we can address the following command to enable our current IP to connect:

```
./mongocli atlas accessLists create --currentIp
```

This should return a result similar to the following:

```
{
  "links": [
    {
      "rel": "self",
      "href": "https://cloud.mongodb.com/api/atlas/v1.0/groups/
[groupid]/accessList?pageNum=1\u0026itemsPerPage=100"
```

```
      }
    ],
    "results": [
      {
        "cidrBlock": "A.B.C.D/32",
        "comment": "My IP Address",
        "groupId": "620173c921b1ab3de3e8e610",
        "ipAddress": "A.B.C.D"
      },
    ],
    "totalCount": 1
}
```

We can create, update, and delete IP blocks that are allowed to connect to MongoDB Atlas and add whole subnets to it.

For example, an entry formatted as 1.2.3.0/24 would allow clients coming from any IP starting with 1.2.3... to connect.

> **Important Note**
> The MongoDB CLI will accept singular and plural naming for the same command. For example, both accessList and accessLists will result in invoking the same command.

Mongo shell and audit logs

We should then install the latest mongosh shell, which replaces the old mongo shell. Similar to the MongoDB CLI, we can install it using either Homebrew or by directly downloading the .zip file and adding mongocli to the shell PATH. Let's get started:

1. Once you have installed the mongosh shell, connecting to MongoDB Atlas is as simple as navigating to the **Databases** dashboard page, clicking on **Connect**, and then clicking on **Connect with the MongoDB Shell**.

2. We can use the connection string by replacing myFirstDatabase with the default database that we entered in the preceding config file, such as the following:

    ```
    $ mongosh "mongodb+srv://mdb52-audit.m5mlj.mongodb.net/
    audit-project" --apiVersion 1 --username admin
    ```

    Doing this will allow us to access the mongosh shell. It will connect to our cloud MongoDB Atlas instance, and use the audit-project database.

In this audit-project database, we have the `contract_signees` collection, which includes the contract information metadata between two or more parties. A sample entry looks like this:

```
{
    _id: ObjectId("6207c023a4edd7e2b6855256"),
    signees: [ 'alex', 'bob', 'catalina' ],
    createdAt: ISODate("2022-02-12T14:11:47.703Z"),
    contractId: '101'
}
```

The `contracts` collection, which includes `contractId` and a link to the URL where we have stored the contract body, looks like this:

```
{
    _id: ObjectId("6207c04ea4edd7e2b6855257"),
    contractId: '101',
    url: 'http://www.example.com/contracts/101'
}
```

Every operation that interacts with the databases has been logged in the audit logs.

3. We can download logs from the MongoDB CLI directly, with the following command:

```
$ mongocli atlas logs download [hostname] [mongodb.gz
|| mongos.gz || mongosqld.gz || mongodb-audit-log.gz ||
mongos-audit-log.gz]
```

We are interested in audit logs, so will use the `mongodb-audit-log.gz` parameter.

4. To find out the [hostname] value, we can use the `process list` command, as follows:

```
$ mongocli atlas process list
```

5. Now, select the `hostname` value or the `userAlias` value from the JSON entry with `typeName: REPLICA_PRIMARY`.

The following code is an example entry for an insert operation:

```
{ "atype" : "authCheck", "ts" : { "$date" : "2022-
02-12T14:11:47.737+00:00" }, "uuid" : { "$binary" :
"2jmqHHr3SoirJJ1/KwHk+Q==", "$type" : "04" }, "local" :
{ "ip" : "192.168.248.220", "port" : 27017 }, "remote"
: { "ip" : "XXX", "port" : 60764 }, "users" : [ {
"user" : "admin", "db" : "admin" } ], "roles" : [ {
"role" : "readWriteAnyDatabase", "db" : "admin" } ],
"param" : { "command" : "insert", "ns" : "audit-project.
```

contract_signees", "args" : { "insert" : "contract_
signees", "documents" : [{ "signees" : ["alex", "bob",
"catalina"], "createdAt" : { "$date" : "2022-02-
12T14:11:47.703+00:00" }, "contractId" : "101", "_id" :
{ "$oid" : "6207c023a4edd7e2b6855256" } }], "ordered" :
true, "apiVersion" : "1", "lsid" : { "id" : { "$binary"
: "rib46IESTA6HDFCbNUwZew==", "$type" : "04" } },
"txnNumber" : { "$numberLong" : "1" }, "$clusterTime" :
{ "clusterTime" : { "$timestamp" : { "t" : 1644674769,
"i" : 1 } }, "signature" : { "hash" : { "$binary" :
"V2BxK73kUbBbObd5o2q3NgbcSXk=", "$type" : "00" }, "keyId"
: { "$numberLong" : "7062257678780727301" } } }, "$db" :
"audit-project" } }, "result" : 0 }

We can witness the inserted document, in bold, inside the args : { } JSON block.

Additionally, the database and collection names are logged in the ns : { } JSON block using the database.collection name convention, as is our IP, which has been replaced by XXX in this example.

6. Since we are logging all operations in our cluster, we can audit and locate an innocuous command such as "read all documents from a collection," as shown in the following example:

{ "atype" : "authCheck", "ts" : { "$date" : "2022-
02-12T14:12:50.773+00:00" }, "uuid" : { "$binary" :
"2jmqHHr3SoirJJ1/KwHk+Q==", "$type" : "04" }, "local" :
{ "ip" : "192.168.248.220", "port" : 27017 }, "remote" :
{ "ip" : "XXX", "port" : 60764 }, "users" : [{ "user"
: "admin", "db" : "admin" }], "roles" : [{ "role" :
"readWriteAnyDatabase", "db" : "admin" }], "param" :
{ "command" : "find", "ns" : "audit-project.contract_
signees", "args" : { "find" : "contract_signees",
"filter" : {}, "apiVersion" : "1", "lsid" : { "id" : {
"$binary" : "rib46IESTA6HDFCbNUwZew==", "$type" : "04" }
}, "$clusterTime" : { "clusterTime" : { "$timestamp" : {
"t" : 1644675168, "i" : 12 } }, "signature" : { "hash" :
{ "$binary" : "vqfrqgLaEJYczWBXED73lp/sXjE=", "$type" :
"00" }, "keyId" : { "$numberLong" : "7062257678780727301"
} } }, "$db" : "audit-project" } }, "result" : 0 }

The preceding audit log line refers to a db.getSiblingDB("audit-project").
contract_signees.find({}) command.

> **Note**
>
> We use `getSiblingDB([database-name])` to denote the database name. If we execute `use audit-project` to connect to the `audit-project` database, then we can skip the `getSiblingDB("audit-project")` part of the command. The command would then be simplified to `db.contract_signees.find({})`.

7. We can also download audit logs via the REST API using the following command:

```
curl --user '[public_key]:[private_key]' --digest \
  --header 'Accept: application/gzip' \
  --request GET "https://cloud.mongodb.com/api/atlas/v1.0/
groups/[project_ID]/clusters/[hostname]/logs/[mongodb.gz
|| mongos.gz || mongosqld.gz || mongodb-audit-log.gz ||
mongos-audit-log.gz]" --output [local_filename]
```

This will download the audit log locally for processing. Additionally, we can set `startDate` and `endDate` in the params to download incrementally. Minimum logs granularity is 5 minutes, and that's the recommended polling interval to download logs.

> **Note**
>
> `group` and `{GROUP-ID}` are both synonymous with the project ID. The project ID is a unique non-human friendly string (an example value is `620173c921b1ab3ee3e8e610`) that MongoDB assigns to each project. To find the value, we can go to our project's home page and then to **...** | **Project Settings**, as shown in the following screenshot:

Figure 4.6 – The Project Settings option

8. Finally, we can review the database access history from the MongoDB Atlas GUI.

Click on the ellipsis button that is on the right-hand side of our project's home page, and then click on **View Database Access History**, as shown in the following screenshot:

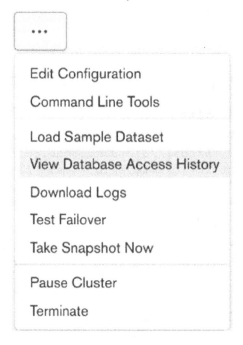

Figure 4.7 – Viewing the database

Clicking on the **View Database Access History** option will get us to a page where we can review audit access requests for the past 7 days by scrolling through them.

> **Important Note**
> This page only shows a small subset of all audit logs – the ones that correspond to database access. We need to refer to the *Audit logging access* section and either download logs or fetch them via the REST API if we need to review all of the operations in the database.

This page is intended as a high-level overview page and not as a substitute for all the other options that we have for accessing and processing audit logs.

Summary

In this chapter, we interrogated the differences between logging and auditing.

Following that, we discussed how to set up auditing, both on-premises using MongoDB Enterprise Edition and in the MongoDB Atlas cloud environment.

Finally, we brought all of the preceding elements together in a case study that uses auditing in the MongoDB Atlas cloud to store, download, and process the audit logs.

We learned how to set up and configure auditing, where it's useful, and how we can store, download, and process the audit logs.

In the next chapter, we will dive deeper into CRUD using three of the most popular languages for web development: Ruby, Python, and PHP.

5

Advanced Querying

In the previous chapter, we learned how to use MongoDB for auditing and the difference in semantics between logging and auditing. In this chapter, we will dive deeper into using MongoDB with drivers and popular frameworks from Ruby, Python, and **Hypertext Preprocessor** (**PHP**).

We will also show the best practices for using these languages, and the variety of comparison and update operators that MongoDB supports at the database level, all of which can be accessed via Ruby, Python, and PHP.

Finally, we will learn about change streams, a MongoDB feature that we can use to stream database changes to our application in real time.

In this chapter, we will cover the following topics:

- MongoDB CRUD operations
- Change streams

Technical requirements

To follow along with the code in this chapter, you need to install MongoDB locally or connect to a MongoDB Atlas database. You can download the MongoDB Community Edition from `mongodb.com` or use the fully managed DBaaS MongoDB Atlas offering, which provides a free tier as well as seamless upgrades to the latest version.

You will also need to download the official drivers for the language of your choice – Ruby, Python, or PHP. You can find all the code from this chapter in this book's GitHub repository at `https://github.com/PacktPublishing/Mastering-MongoDB-6.x`.

MongoDB CRUD operations

In this section, we will cover CRUD operations while using Ruby, Python, and PHP with the official MongoDB driver and some popular frameworks for each language, respectively.

CRUD using the Ruby driver

In *Chapter 3, MongoDB CRUD Operations*, we covered how to connect to MongoDB from Ruby, Python, and PHP using the respective drivers and **object document mapping (ODM)**. In this chapter, we will explore the `create`, `read`, `update`, and `delete` operations while using the official drivers and the most commonly used ODM frameworks.

Creating documents

Using the process described in *Chapter 2, Schema Design and Data Modeling*, we assume that we have an `@collection` instance variable pointing to our `books` collection in a `mongo_book` database in the `127.0.0.1:27017` default database:

```
@collection = Mongo::Client.new([ '127.0.0.1:27017' ],
:database => 'mongo_book').database[:books]
```

Let's insert a single document with our definition, as follows:

```
document = { isbn: '101', name: 'Mastering MongoDB', price: 30}
```

This can be performed with a single line of code, as follows:

```
result = @collection.insert_one(document)
```

The resulting object is a `Mongo::Operation::Result` class with content that is similar to what we had in the shell, as shown in the following code:

```
{"n"=>1, "ok"=>1.0}
```

Here, n is the number of affected documents; 1 means we inserted one object, while ok means 1 (`true`).

Creating multiple documents in one step is similar to this. For two documents with `isbn` values of `102` and `103`, and using `insert_many` instead of `insert_one`, we have the following code:

```
documents = [ { isbn: '102', name: 'MongoDB in 7 years', price:
50 },
                { isbn: '103', name: 'MongoDB for experts', price:
40 } ]
result = @collection.insert_many(documents)
```

The resulting object is now a `Mongo::BulkWrite::Result` class, meaning that the `BulkWrite` interface was used for improved performance.

The main difference is that we now have an attribute, `inserted_ids,`, which will return `ObjectId` of the inserted objects from the `BSON::ObjectId` class.

Reading data

Finding documents works in the same way as creating them – that is, at the collection level:

```
@collection.find( { isbn: '101' } )
```

Multiple search criteria can be chained and are equivalent to the AND operator in SQL:

```
@collection.find( { isbn: '101', name: 'Mastering MongoDB' } )
```

The `mongo-ruby-driver` API provides several query options to enhance queries; the most widely used query options are listed in the following table:

Option	Description
`allow_partial_results`	This is for use with sharded clusters. If a shard is down, it allows the query to return results from the shards that are up, potentially getting only a portion of the results.
`batch_size(Integer)`	This can change the batch size that the cursor will fetch from MongoDB. This is done on each `GETMORE` operation (for example, by typing it on the mongo shell).
`comment(String)`	With this command, we can add a comment in our query for documentation reasons.
`hint(Hash)`	We can force usage of an index using `hint()`.
`limit(Integer)`	We can limit the result set to the number of documents specified by `Integer`.
`max_scan(Integer)`	We can limit the number of documents that will be scanned. This will return incomplete results and is useful if we are performing operations where we want to guarantee that they won't take a long time, such as when we connect to our production database.
`no_cursor_timeout`	If we don't specify this parameter, MongoDB will close any inactive cursor after 600 seconds. With this parameter, our cursor will never be closed.
`projection(Hash)`	We can use this parameter to fetch or exclude specific attributes from our results. This will reduce data transfer over the wire. An example of this is `client[:books].find.projection(:price => 1)`.
`read(Hash)`	We can specify a read preference to be applied only for this query: `client[:books].find.read(:mode => :secondary_preferred)`.
`show_disk_loc(Boolean)`	We should use this option if we want to find the actual location of our results on a disk.
`skip(Integer)`	This can be used to skip the specified number of documents. It's useful for paginating results.
`snapshot`	This can be used to execute our query in snapshot mode. This is useful when we want a more stringent consistency.
`sort(Hash)`	We can use this to sort our results; for example, `client[:books].find.sort(:name => -1)`.

Table 5.1 – MongoDB Ruby client query options

On top of the query options, `mongo-ruby-driver` provides some helper functions that can be chained at the method call level, as follows:

- `.count`: The total count for the preceding query

- `.distinct(:field_name)`: To distinguish between the results of the preceding query by `:field_name`

`find()` returns a cursor containing the result set that we can iterate using `.each` in Ruby, such as every other object:

```
result = @collection.find({ isbn: '101' })
result.each do |doc|
  puts doc.inspect
end
```

The output for our `books` collection is as follows:

```
{"_id"=>BSON::ObjectId('592149c4aabac953a3a1e31e'),
"isbn"=>"101", "name"=>"Mastering MongoDB", "price"=>30.0,
"published"=>2017-06-25 00:00:00 UTC}
```

Chaining operations in find()

`find()`, by default, uses an AND operator to match multiple fields. If we want to use an OR operator, our query needs to be as follows:

```
result = @collection.find('$or' => [{ isbn: '101' }, { isbn:
'102' }]).to_a
puts result
```

The output of the preceding code is as follows:

```
{"_id"=>BSON::ObjectId('592149c4aabac953a3a1e31e'),
"isbn"=>"101", "name"=>"Mastering MongoDB",
"price"=>30.0, "published"=>2017-06-25 00:00:00 UTC}
{"_id"=>BSON::ObjectId('59214bc1aabac954263b24e0'),
"isbn"=>"102", "name"=>"MongoDB in 7 years", "price"=>50.0,
"published"=>2017-06-26 00:00:00 UTC}
```

We can also use `$and` instead of `$or` in the previous example:

```
result = @collection.find('$and' => [{ isbn: '101' }, { isbn:
'102' }]).to_a
puts result
```

This, of course, will return no results since no document can have isbn values of 101 and 102.

An interesting and hard bug to find is if we define the same key multiple times, such as in the following code:

```
result = @collection.find({ isbn: '101', isbn: '102' })
puts result
```

The output of the preceding code is as follows:

```
{"_id"=>BSON::ObjectId('59214bc1aabac954263b24e0'),
"isbn"=>"102", "name"=>"MongoDB in 7 years", "price"=>50.0,
"published"=>2017-06-26 00:00:00 UTC}
```

In comparison, the opposite order will cause the document with an isbn value of 101 to be returned:

```
result = @collection.find({ isbn: '102', isbn: '101' })
puts result
```

The output of the preceding code is as follows:

```
{"_id"=>BSON::ObjectId('592149c4aabac953a3a1e31e'),
"isbn"=>"101", "name"=>"Mastering MongoDB", "price"=>30.0,
"published"=>2017-06-25 00:00:00 UTC}
```

> **Note**
>
> This is because in Ruby hashes, by default, all duplicated keys except for the last one are silently ignored. This may not happen in the simplistic form shown in the preceding example, but it is prone to happen if we create keys programmatically.

Nested operations

Accessing embedded documents in mongo-ruby-driver is as simple as using the dot notation:

```
result = @collection.find({'meta.authors': 'alex giamas'}).to_a
puts result
```

The output of the preceding code is as follows:

```
"_id"=>BSON::ObjectId('593c24443c8ca55b969c4c54'),
"isbn"=>"201", "name"=>"Mastering MongoDB, 3rd Edition",
"meta"=>{"authors"=>"alex giamas"}}
```

> **Note**
>
> We need to enclose the key name in quotes (' ') to access the embedded object, just as we need it for operations starting with $, such as '$set'.

Updating data

Updating documents using mongo-ruby-driver is chained to finding them. Using our example books collection, we can do the following:

```
@collection.update_one( { 'isbn': 101}, { '$set' => { name:
'Mastering MongoDB, 3rd Edition' } } )
```

This finds the document with an isbn value of 101 and changes its name to Mastering MongoDB, 3rd Edition.

In a similar way to update_one, we can use update_many to update multiple documents retrieved via the first parameter of the method.

> **Note**
>
> If we don't use the $set operator, the contents of the document will be replaced by the new document.

With any Ruby version 2.2 or later, keys can be either quoted or unquoted; however, keys that start with $ need to be quoted as follows:

```
@collection.update( { isbn: '101'}, { "$set": { name:
"Mastering MongoDB, 3rd edition" } } )
```

The resulting object of an update will contain information about the operation, including the following methods:

- ok?: A Boolean value that shows whether the operation was successful or not
- matched_count: The number of documents matching the query
- modified_count: The number of documents affected (updated)
- upserted_count: The number of documents upserted if the operation includes $set
- upserted_id: The unique ObjectId of the upserted document if there is one

Updates that modify fields of a constant data size will be *in place*; this means that they won't move the document from its physical location on the disk. This includes operations such as $inc and $set on the Integer and Date fields.

Updates that can increase the size of a document may result in the document being moved from its physical location on the disk to a new location at the end of the file. In this case, queries may miss or return the document multiple times. To avoid this, we can use $snapshot: true while querying.

Deleting data

Deleting documents works in a similar way to finding documents. We need to find documents and then apply the delete operation.

For example, with our books collection, which we used previously, we can issue the following code:

```
@collection.find( { isbn: '101' } ).delete_one
```

This will delete a single document. In our case, since isbn is unique for every document, this is expected. If our find() clause had matched multiple documents, then delete_one would have deleted just the first one that find() returned, which may or may not have been what we wanted.

> **Note**
>
> If we use delete_one with a query matching multiple documents, the results may be unexpected.

If we want to delete all documents matching our find() query, we have to use delete_many, as follows:

```
@collection.find( { price: { $gte: 30 } ).delete_many
```

In the preceding example, we are deleting all books that have a price greater than or equal to 30.

Batch operations

We can use the BulkWrite API for batch operations. In our previous *insert many documents* example in the first section of this chapter, *Creating documents*, this would be as follows:

```
@collection.bulk_write([ { insertMany: documents
                      }],
                  ordered: true)
```

The BulkWrite API can take the following parameters:

- insertOne
- updateOne
- updateMany

- `replaceOne`

- `deleteOne`

- `deleteMany`

One version of these commands will `insert/update/replace/delete` a single document, even if the filter that we specify matches more than one document. In this case, it's important to have a filter that matches a single document to avoid unexpected behaviors.

It's also possible, and a perfectly valid use case, to include several operations in the first argument of the `bulk_write` command. This allows us to issue commands in a sequence when we have operations that depend on each other, and we want to batch them in a logical order according to our business logic. Any error will stop `ordered:true` batch writes, and we will need to manually roll back our operations. A notable exception is `writeConcern` errors – for example, requesting a majority of our replica set members to acknowledge our write. In this case, batch writes will go through, and we can observe the errors in the `writeConcernErrors` result field:

```
old_book = @collection.findOne(name: 'MongoDB for experts')
new_book = { isbn: 201, name: 'MongoDB for experts, 3rd
Edition', price: 55 }
@collection.bulk_write([ {deleteOne: old_book}, { insertOne:
new_book
                    }],
             ordered: true)
```

In the previous example, we made sure that we deleted the original book before adding the new (and more expensive) edition of our `MongoDB for experts` book.

BulkWrite can batch up to 1,000 operations. If we have more than 1,000 underlying operations in our commands, these will be split into chunks of thousands. It is good practice to try to keep our write operations to a single batch if we can, to avoid unexpected behavior.

CRUD in Mongoid

In this section, we will use Mongoid to perform `create`, `read`, `update`, and `delete` operations. All of this code is also available on GitHub at `https://github.com/PacktPublishing/Mastering-MongoDB-6.x/tree/main/chapter_5`.

Reading data

Back in *Chapter 2*, *Schema Design and Data Modeling*, we described how to install, connect, and set up models, including inheritance, in Mongoid. Here, we will go through the most common use cases of CRUD.

Finding documents is done using a DSL similar to **Active Record** (**AR**). As with AR, where a relational database is used, Mongoid assigns a class to a MongoDB collection (a table) and any object instance to a document (a row from a relational database):

```
Book.find('592149c4aabac953a3a1e31e')
```

This will find the document by ObjectId and return the document with isbn 101, as will the query by a name attribute:

```
Book.where(name: 'Mastering MongoDB')
```

In a similar fashion to the dynamically generated AR queries by an attribute, we can use the helper method:

```
Book.find_by(name: 'Mastering MongoDB')
```

This queries by attribute name, which is equivalent to the previous query.

We should enable QueryCache to avoid hitting the database for the same query multiple times, as follows:

```
Mongoid::QueryCache.enabled = true
```

This can be added to any code block that we want to enable, or to the initializer for Mongoid.

Scoping queries

We can scope queries in Mongoid using class methods, as follows:

```
Class Book
...
  def self.premium
    where(price: {'$gt': 20'})
  end
End
```

Then, we can use the following query:

```
Book.premium
```

This will query for books with a price greater than 20.

Create, update, and delete

The Ruby interface for creating documents is similar to an AR:

```
Book.where(isbn: 202, name: 'Mastering MongoDB, 3rd Edition').
create
```

This will return an error if the creation fails.

We can use the bang version to force an exception to be raised if saving the document fails:

```
Book.where(isbn: 202, name: 'Mastering MongoDB, 3rd Edition').
create!
```

The BulkWrite API is not supported as of Mongoid version 6.x. The workaround is to use the mongo-ruby-driver API, which will not use the `mongoid.yml` configuration or custom validations. Otherwise, you can use `insert_many([array_of_documents])`, which will insert the documents one by one.

To update documents, we can use `update` or `update_all`. Using `update` will update only the first document retrieved by the query part, whereas `update_all` will update all of them:

```
Book.where(isbn: 202).update(name: 'Mastering MongoDB, 3rd
Edition')
Book.where(price: { '$gt': 20 }).update_all(price_range:
'premium')
```

Deleting a document is similar to creating it – we provide `delete` to skip callbacks and `destroy` if we want to execute any available callbacks in the affected document.

`delete_all` and `destroy_all` are convenient methods for multiple documents.

> **Note**
>
> `destroy_all` should be avoided, if possible, as it will load all documents into memory to execute callbacks and thus can be memory-intensive.

CRUD using the Python driver

PyMongo is the officially supported driver for Python by MongoDB. In this section, we will use PyMongo to `create`, `read`, `update`, and `delete` documents in MongoDB.

Creating and deleting data

The Python driver provides methods for CRUD just like Ruby and PHP. Following on from *Chapter 2*, *Schema Design and Data Modeling*, and the `books` variable that points to our `books` collection, we will write the following code block:

```python
from pymongo import MongoClient
from pprint import pprint

>>> book = {
 'isbn': '301',
 'name': 'Python and MongoDB',
 'price': 60
}
>>> insert_result = books.insert_one(book)
>>> pprint(insert_result)

<pymongo.results.InsertOneResult object at 0x104bf3370>

>>> result = list(books.find())
>>> pprint(result)

[{u'_id': ObjectId('592149c4aabac953a3a1e31e'),
 u'isbn': u'101',
 u'name': u'Mastering MongoDB',
 u'price': 30.0,
 u'published': datetime.datetime(2017, 6, 25, 0, 0)},
{u'_id': ObjectId('59214bc1aabac954263b24e0'),
 u'isbn': u'102',
 u'name': u'MongoDB in 7 years',
 u'price': 50.0,
 u'published': datetime.datetime(2017, 6, 26, 0, 0)},
{u'_id': ObjectId('593c24443c8ca55b969c4c54'),
 u'isbn': u'201',
 u'meta': {u'authors': u'alex giamas'},
 u'name': u'Mastering MongoDB, 3rd Edition'},
{u'_id': ObjectId('594061a9aabac94b7c858d3d'),
```

```
u'isbn': u'301',
u'name': u'Python and MongoDB',
u'price': 60}]
```

In the previous example, we used `insert_one()` to insert a single document, which we can define using the Python dictionary notation; we can then query it for all the documents in the collection.

The resulting object for `insert_one` and `insert_many` has two fields of interest:

- `Acknowledged`: A Boolean that is `true` if the insert has succeeded and `false` if it hasn't, or if the write concern is 0 (a *fire and forget* write).
- `inserted_id` for `insert_one`: The `ObjectId` property of the written document and the `inserted_id` properties for `insert_many`. This is the array of ObjectIds of the written documents.

We used the `pprint` library to pretty-print the `find()` results. The built-in way to iterate through the result set is by using the following code:

```
for document in results:
    print(document)
```

Deleting documents works in a similar way to creating them. We can use `delete_one` to delete the first instance or `delete_many` to delete all instances of the matched query:

```
>>> result = books.delete_many({ "isbn": "101" })
>>> print(result.deleted_count)
1
```

The `deleted_count` instance tells us how many documents were deleted; in our case, it is 1, even though we used the `delete_many` method.

To delete all documents from a collection, we can pass in the empty document, `{}`.

To drop a collection, we can use `drop()`:

```
>>> books.delete_many({})
>>> books.drop()
```

Finding documents

To find documents based on top-level attributes, we can simply use a dictionary:

```
>>> books.find({"name": "Mastering MongoDB"})

[{u'_id': ObjectId('592149c4aabac953a3a1e31e'),
```

```
    u'isbn': u'101',
    u'name': u'Mastering MongoDB',
    u'price': 30.0,
    u'published': datetime.datetime(2017, 6, 25, 0, 0)}]
```

To find documents in an embedded document, we can use dot notation. In the following example, we are using meta.authors to access the authors embedded document inside the meta document:

```
>>> result = list(books.find({"meta.authors": {"$regex":
"aLEx", "$options": "i"}}))
>>> pprint(result)


[{u'_id': ObjectId('593c24443c8ca55b969c4c54'),
  u'isbn': u'201',
  u'meta': {u'authors': u'alex giamas'},
  u'name': u'Mastering MongoDB, 3rd Edition'}]
```

In this example, we used a regular expression to match aLEx, which is case insensitive, in every document where the string is mentioned in the meta.authors embedded document. PyMongo uses this notation for regular expression queries, called the $regex notation in MongoDB documentation. The second parameter is the options parameter for $regex, which we will explain in detail in the *Using regular expressions* section later in this chapter.

Comparison operators are also supported, and a full list of these is given in the *Comparison operators* section, later in this chapter:

```
>>> result = list(books.find({ "price": {   "$gt":40 } }))
>>> pprint(result)


[{u'_id': ObjectId('594061a9aabac94b7c858d3d'),
  u'isbn': u'301',
  u'name': u'Python and MongoDB',
  u'price': 60}]
```

Let's add multiple dictionaries to our query results in a logical AND query:

```
>>> result = list(books.find({"name": "Mastering MongoDB",
"isbn": "101"}))
>>> pprint(result)


[{u'_id': ObjectId('592149c4aabac953a3a1e31e'),
```

```
u'isbn': u'101',
u'name': u'Mastering MongoDB',
u'price': 30.0,
u'published': datetime.datetime(2017, 6, 25, 0, 0)}]
```

For books that have both isbn=101 and name=Mastering MongoDB, to use logical operators such as $or and $and, we must use the following syntax:

```
>>> result = list(books.find({"$or": [{"isbn": "101"}, {"isbn":
"102"}]}))
>>> pprint(result)

[{u'_id': ObjectId('592149c4aabac953a3a1e31e'),
 u'isbn': u'101',
 u'name': u'Mastering MongoDB',
 u'price': 30.0,
 u'published': datetime.datetime(2017, 6, 25, 0, 0)},
{u'_id': ObjectId('59214bc1aabac954263b24e0'),
 u'isbn': u'102',
 u'name': u'MongoDB in 7 years',
 u'price': 50.0,
 u'published': datetime.datetime(2017, 6, 26, 0, 0)}]
```

For books that have an isbn value of 101 or 102, if we want to combine the AND and OR operators, we must use the $and operator, as follows:

```
>>> result = list(books.find({"$or": [{"$and": [{"name":
"Mastering MongoDB", "isbn": "101"}]}, {"$and": [{"name":
"MongoDB in 7 years", "isbn": "102"}]}]}))
>>> pprint(result)
[{u'_id': ObjectId('592149c4aabac953a3a1e31e'),
 u'isbn': u'101',
 u'name': u'Mastering MongoDB',
 u'price': 30.0,
 u'published': datetime.datetime(2017, 6, 25, 0, 0)},
{u'_id': ObjectId('59214bc1aabac954263b24e0'),
 u'isbn': u'102',
 u'name': u'MongoDB in 7 years',
 u'price': 50.0,
 u'published': datetime.datetime(2017, 6, 26, 0, 0)}]
```

For a result of OR between two queries, consider the following:

- The first query is asking for documents that have isbn=101 AND name=Mastering MongoDB

- The second query is asking for documents that have isbn=102 AND name=MongoDB in 7 years

- The result is the union of these two datasets

Updating documents

In the following code block, you can see an example of updating a single document using the update_one helper method.

This operation matches one document in the search phase and modifies one document based on the operation to be applied to the matched documents:

```
>>> result = books.update_one({"isbn": "101"}, {"$set":
{"price": 100}})
>>> print(result.matched_count)
1
>>> print(result.modified_count)
1
```

In a similar way to inserting documents, when updating documents, we can use update_one or update_many:

- The first argument here is the filter document for matching the documents that will be updated

- The second argument is the operation to be applied to the matched documents

- The third (optional) argument is to use upsert=false (the default) or true, which is used to create a new document if it's not found

Another interesting argument is bypass_document_validation=false (the default) or true, which is optional. This will ignore validations (if there are any) for the documents in the collection.

The resulting object will have matched_count for the number of documents that matched the filter query, and modified_count for the number of documents that were affected by the update part of the query.

In our example, we are setting price=100 for the first book with isbn=101 through the $set update operator. A list of all update operators can be found in the *Update operators* section later in this chapter.

> **Note**
>
> If we don't use an `update` operator as the second argument, the contents of the matched document will be entirely replaced by the new document.

CRUD using PyMODM

PyMODM is a core ODM that provides simple and extensible functionality. It is developed and maintained by MongoDB's engineers who get fast updates and support for the latest stable version of MongoDB available.

In *Chapter 2, Schema Design and Data Modeling*, we explored how to define different models and connect to MongoDB. CRUD, when using PyMODM, as with every ODM, is simpler than when using low-level drivers.

Creating documents

A new `user` object, as defined in *Chapter 2, Schema Design and Data Modeling*, can be created with a single line:

```
>>> user = User('alexgiamas@packt.com', 'Alex', 'Giamas').
save()
```

In this example, we used positional arguments in the same order that they were defined in the `user` model to assign values to the `user` model attributes.

We can also use keyword arguments or a mix of both, as follows:

```
>>> user = User(email='alexgiamas@packt.com', 'Alex', last_
name='Giamas').save()
```

Bulk saving can be done by passing in an array of users to `bulk_create()`:

```
>>> users = [ user1, user2,...,userN]
>>>   User.bulk_create(users)
```

Updating documents

We can modify a document by directly accessing the attributes and calling `save()` again:

```
>>> user.first_name = 'Alexandros'
>>> user.save()
```

If we want to update one or more documents, we must use `raw()` to filter out the documents that will be affected and chain `update()` to set the new values:

```
>>> User.objects.raw({'first_name': {'$exists': True}})
              .update({'$set': {'updated_at': datetime.
datetime.now()}})
```

In the preceding example, we search for all `User` documents that have a first name and set a new field, `updated_at`, to the current timestamp. The result of the `raw()` method is `QuerySet`, a class used in PyMODM to handle queries and work with documents in bulk.

Deleting documents

Deleting an API is similar to updating it – by using `QuerySet` to find the affected documents and then chaining on a `.delete()` method to delete them:

```
>>> User.objects.raw({'first_name': {'$exists': True}}).
delete()
```

Querying documents

Querying is done using `QuerySet`, as described previously.

Some of the convenience methods that are available include the following:

- `all()`
- `count()`
- `first()`
- `exclude(*fields)`: To exclude some fields from the result
- `only(*fields)`: To include only some fields in the result (this can be chained for a union of fields)
- `limit(limit)`
- `order_by(ordering)`
- `reverse()`: If we want to reverse the `order_by()` order
- `skip(number)`
- `values()`: To return Python `dict` instances instead of model instances

By using `raw()`, we can use the same queries that we described in the previous PyMongo section for querying and still exploit the flexibility and convenience methods provided by the ODM layer.

CRUD using the PHP driver

In PHP, we should use the `mongo-php-library` driver instead of the deprecated MongoClient. The overall architecture was explained in *Chapter 2*, *Schema Design and Data Modeling*. Here, we will cover more details regarding the API and how we can perform CRUD operations using it.

Creating and deleting data

The following command will insert a single `$document` that contains an array of two key/value pairs, with the key names of `isbn` and `name`:

```php
$document = array( "isbn" => "401", "name" => "MongoDB and PHP"
);
$result = $collection->insertOne($document);
var_dump($result);
```

The output from the `var_dump($result)` command is as follows:

```
MongoDB\InsertOneResult Object
(
    [writeResult:MongoDB\InsertOneResult:private] => MongoDB\
Driver\WriteResult Object
        (
            [nInserted] => 1
            [nMatched] => 0
            [nModified] => 0
            [nRemoved] => 0
            [nUpserted] => 0
            [upsertedIds] => Array
                (
                )

            [writeErrors] => Array
                (
                )

            [writeConcernError] =>
            [writeConcern] => MongoDB\Driver\WriteConcern Object
                (
                )
```

```
        )


    [insertedId:MongoDB\InsertOneResult:private] => MongoDB\
BSON\ObjectID Object
        (
                [oid] => 5941ac50aabac9d16f6da142
        )


    [isAcknowledged:MongoDB\InsertOneResult:private] => 1
)
```

This rather lengthy output contains all the information that we may need. Here, we can insert the ObjectId property of the document, the number of inserted, matched, modified, removed, and upserted documents by fields prefixed with n, and information about writeError or writeConcernError.

There are also convenience methods in the $result object if we want to get this information:

- $result->getInsertedCount(): To get the number of inserted objects
- $result->getInsertedId(): To get the ObjectId property of the inserted document

We can also use the ->insertMany() method to insert many documents at once, as follows:

```
$documentAlpha = array( "isbn" => "402", "name" => "MongoDB and
PHP, 3rd Edition" );
$documentBeta  = array( "isbn" => "403", "name" => "MongoDB and
PHP, revisited" );
$result = $collection->insertMany([$documentAlpha,
$documentBeta]);

print_r($result);
```

The result is as follows:

```
(
    [writeResult:MongoDB\InsertManyResult:private] => MongoDB\
Driver\WriteResult Object
        (
                [nInserted] => 2
                [nMatched] => 0
```

```
            [nModified] => 0
            [nRemoved] => 0
            [nUpserted] => 0
            [upsertedIds] => Array
               (
               )

            [writeErrors] => Array
               (
               )

            [writeConcernError] =>
            [writeConcern] => MongoDB\Driver\WriteConcern Object
               (
               )

        )

    [insertedIds:MongoDB\InsertManyResult:private] => Array
        (
            [0] => MongoDB\BSON\ObjectID Object
                (
                    [oid] => 5941ae85aabac9d1d16c63a2
                )

            [1] => MongoDB\BSON\ObjectID Object
                (
                    [oid] => 5941ae85aabac9d1d16c63a3
                )

        )

    [isAcknowledged:MongoDB\InsertManyResult:private] => 1
)
```

Again, `$result->getInsertedCount()` will return 2, whereas `$result->getInsertedIds()` will return an array with the two newly created `ObjectId` properties:

```
array(2) {
  [0]=>
  object(MongoDB\BSON\ObjectID)#13 (1) {
    ["oid"]=>
    string(24) "5941ae85aabac9d1d16c63a2"
  }
  [1]=>
  object(MongoDB\BSON\ObjectID)#14 (1) {
    ["oid"]=>
    string(24) "5941ae85aabac9d1d16c63a3"
  }
}
```

Deleting documents is a similar process to inserting documents, but it uses the `deleteOne()` and `deleteMany()` methods instead. An example of `deleteMany()` is as follows:

```
$deleteQuery = array( "isbn" => "401");
$deleteResult = $collection->deleteMany($deleteQuery);
print($deleteResult->getDeletedCount());
```

The following code block shows the output:

```
MongoDB\DeleteResult Object
(
    [writeResult:MongoDB\DeleteResult:private] => MongoDB\
Driver\WriteResult Object
        (
            [nInserted] => 0
            [nMatched] => 0
            [nModified] => 0
            [nRemoved] => 2
            [nUpserted] => 0
            [upsertedIds] => Array
                (
                )
```

```
                    [writeErrors] => Array
                        (
                        )


                    [writeConcernError] =>
                    [writeConcern] => MongoDB\Driver\WriteConcern Object
                        (
                        )



            )


        [isAcknowledged:MongoDB\DeleteResult:private] => 1
    )
    2
```

In this example, we used `->getDeletedCount()` to get the number of affected documents, which is printed in the last line of the output.

BulkWrite

The new PHP driver supports the `BulkWrite` interface to minimize network calls to MongoDB:

```php
$manager = new MongoDB\Driver\Manager('mongodb://
localhost:27017');
$bulk = new MongoDB\Driver\BulkWrite(array("ordered" => true));
$bulk->insert(array( "isbn" => "401", "name" => "MongoDB and
PHP" ));
$bulk->insert(array( "isbn" => "402", "name" => "MongoDB and
PHP, 3rd Edition" ));
$bulk->update(array("isbn" => "402"), array('$set' =>
array("price" => 15)));
$bulk->insert(array( "isbn" => "403", "name" => "MongoDB and
PHP, revisited" ));


$result = $manager->executeBulkWrite('mongo_book.books',
$bulk);
print_r($result);
```

The result is as follows:

```
MongoDB\Driver\WriteResult Object
(
    [nInserted] => 3
    [nMatched] => 1
    [nModified] => 1
    [nRemoved] => 0
    [nUpserted] => 0
    [upsertedIds] => Array
        (
        )


    [writeErrors] => Array
        (
        )


    [writeConcernError] =>
    [writeConcern] => MongoDB\Driver\WriteConcern Object
        (
        )


)
```

In the preceding example, we executed two inserts, one `update`, and a third `insert` in an ordered fashion. The `WriteResult` object contains a total of three inserted documents and one modified document.

The main difference compared to simple create/delete queries is that `executeBulkWrite()` is a method of the `MongoDB\Driver\Manager` class, which we instantiate on the first line.

Read

Querying an interface is similar to inserting and deleting, with the `findOne()` and `find()` methods used to retrieve the first result or all results of a query:

```
$document = $collection->findOne( array("isbn" => "401") );
$cursor = $collection->find( array( "name" => new MongoDB\BSON\
Regex("mongo", "i") ) );
```

In the second example, we are using a regular expression to search for a key name with the value, mongo (which is case insensitive).

Embedded documents can be queried using the . notation, as with the other languages that we examined earlier in this chapter:

```
$cursor = $collection->find( array('meta.price' => 50) );
```

We are doing this to query for a price embedded document inside the meta key field.

Similar to Ruby and Python, in PHP, we can query using comparison operators, as shown in the following code:

```
$cursor = $collection->find( array( 'price' => array('$gte'=>
60) ) );
```

A complete list of comparison operators supported in the PHP driver can be found at the end of this chapter.

Querying with multiple key-value pairs is an implicit AND, whereas queries that use $or, $in, $nin, or AND ($and) combined with $or can be achieved with nested queries:

```
$cursor = $collection->find( array( '$or' => array(
                                        array("price" =>
array( '$gte' => 60)),
                                        array("price" =>
array( '$lte' => 20))
                                    )));
```

This finds documents that have price>=60 OR price<=20.

Updating documents

Updating documents has a similar interface to the ->updateOne() OR ->updateMany() method.

The first parameter is the query to find documents; the second one will update our documents.

We can use any of the update operators explained at the end of this chapter to update in place, or specify a new document to completely replace the document in the query:

```
$result = $collection->updateOne(
   array( "isbn" => "401"),
    array( '$set' => array( "price" => 39 ) )
);
```

> **Note**
>
> We can use single quotes or double quotes for key names, but if we have special operators starting with $, we need to use single quotes. We can use `array( "key" => "value" )` or `["key" => "value"]`. We prefer the more explicit `array()` notation in this book.

The `->getMatchedCount()` and `->getModifiedCount()` methods will return the number of documents matched in the query part, or the ones modified from the query. If the new value is the same as the existing value of a document, it will not be counted as modified.

CRUD using Doctrine

Following on from our Doctrine example in *Chapter 2*, *Schema Design and Data Modeling*, we will work on these models for CRUD operations.

Creating, updating, and deleting

Creating documents is a two-step process. First, we must create our document and set the attribute values:

```
$book = new Book();
$book->setName('MongoDB with Doctrine');
$book->setPrice(45);
```

Following this, we must ask Doctrine to save $book in the next `flush()` call:

```
$dm->persist($book);
```

We can force save by manually calling `flush()`, as follows:

```
$dm->flush();
```

In this example, $dm is a `DocumentManager` object that we use to connect to our MongoDB instance, as follows:

```
$dm = DocumentManager::create(new Connection(), $config);
```

Updating a document is as easy as assigning values to the attributes:

```
$book->price = 39;
$book->persist($book);
```

This will save our `MongoDB with Doctrine` book with the new price of 39.

Updating documents in place uses the `QueryBuilder` interface.

Doctrine provides several helper methods around atomic updates, as follows:

- `set($name, $value, $atomic = true)`
- `setNewObj($newObj)`
- `inc($name, $value)`
- `unsetField($field)`
- `push($field, $value)`
- `pushAll($field, array $valueArray)`
- `addToSet($field, $value)`
- `addManyToSet($field, array $values)`
- `popFirst($field)`
- `popLast($field)`
- `pull($field, $value)`
- `pullAll($field, array $valueArray)`

`update` will, by default, update the first document found by the query. If we want to change multiple documents, we need to use `->updateMany()`:

```
$dm->createQueryBuilder('Book')
    ->updateMany()
    ->field('price')->set(69)
    ->field('name')->equals('MongoDB with Doctrine')
    ->getQuery()
    ->execute();
```

In the preceding example, we are setting the price of the book to 69 with name=`'MongoDB with Doctrine'`. The list of comparison operators in Doctrine is available in the following *Read* section.

We can chain multiple comparison operators, resulting in an AND query and also multiple helper methods, resulting in updates to several fields.

Deleting a document is similar to creating it, as shown in the following code block:

```
$dm->remove($book);
```

Removing multiple documents is best done using the `QueryBuilder` interface, which we will explore further in the following section:

```
$qb = $dm->createQueryBuilder('Book');
$qb->remove()
```

```
->field('price')->equals(50)
->getQuery()
->execute();
```

Read

Doctrine provides a `QueryBuilder` interface for building queries for MongoDB. Given that we have defined our models as described in *Chapter 2, Schema Design and Data Modeling*, we can do this to obtain an instance of a `QueryBuilder` interface named $db, get a default find-all query, and execute it, as follows:

```
$qb = $dm->createQueryBuilder('Book');
$query = $qb->getQuery();
$books = $query->execute();
```

The $books variable now contains an iterable lazy data-loading cursor over our result set.

Using $qb->eagerCursor(true); over the `QueryBuilder` object will return an eager cursor that fetches all the data from MongoDB as soon as we start iterating our results.

Some helper methods for querying are as follows:

- ->getSingleResult(): This is equivalent to findOne().
- ->select('name'): This returns only the values for the 'key' attribute from our books collection. ObjectId will always be returned.
- ->hint('book_name_idx'): This forces the query to use this index. We'll see more about indexes in *Chapter 7, Aggregation*.
- ->distinct('name'): This returns distinct results by name.
- ->limit(10): This returns the first 10 results.
- ->sort('name', 'desc'): This sorts by name (such as desc or asc).

Doctrine uses the concept of hydration when fetching documents from MongoDB. Hydration defines the query's result schema. We can, for example, configure hydration to return a collection of objects, a single scalar value, or an array of arrays representing different records. Using an identity map, it will cache MongoDB results in memory and consult this map before hitting the database. Disabling hydration can be done per query by using ->hydration(false) or globally using the configuration, as explained in *Chapter 2, Schema Design and Data Modeling*.

We can also force Doctrine to refresh data in the identity map for a query from MongoDB using ->refresh() on $qb.

The comparison operators that we can use with Doctrine are as follows:

- `where($javascript)`
- `in($values)`
- `notIn($values)`
- `equals($value)`
- `notEqual($value)`
- `gt($value)`
- `gte($value)`
- `lt($value)`
- `lte($value)`
- `range($start, $end)`
- `size($size)`
- `exists($bool)`
- `type($type)`
- `all($values)`
- `mod($mod)`
- `addOr($expr)`
- `addAnd($expr)`
- `references($document)`
- `includesReferenceTo($document)`

Consider the following query as an example:

```
$qb = $dm->createQueryBuilder('Book')
            ->field('price')->lt(30);
```

This will return all books whose price is less than 30.

`addAnd()` may seem redundant since chaining multiple query expressions in Doctrine is implicitly AND, but it is useful if we want to do `AND ( (A OR B), (C OR D) )`, where A, B, C, and D are standalone expressions.

To nest multiple OR operators with an external AND query, and in other equally complex cases, the nested ORs need to be evaluated as expressions using ->expr():

```
$expression = $qb->expr()->field('name')->equals('MongoDB with
Doctrine')
```

$expression is a standalone expression that can be used with $qb->addOr($expression) and similarly with addAnd().

Best practices

Some best practices for using Doctrine with MongoDB are as follows:

- Don't use unnecessary cascading.
- Don't use unnecessary life cycle events.
- Don't use special characters such as non-ASCII ones in class, field, table, or column names, as Doctrine is not Unicode-safe yet.
- Initialize collection references in the model's constructor. Constrain relationships between objects as much as possible.
- Avoid bidirectional associations between models and eliminate the ones that are not needed. This helps with performance and loose coupling and produces simpler and more easily maintainable code.

The following is a list of all comparison operators that MongoDB supports as of version 6:

Name	Description
$eq	Will match with all test values that are equal to the target value
$gt	Will match with all test values that are greater than the target value
$gte	Will match with all test values that are greater than or equal to the target value
$lt	Will match with all test values that are less than the target value
$lte	Will match with all test values that are less than or equal to the target value
$ne	Will match with all test values that are NOT equal to the target value
$in	Will match the target value with ANY of the test values in the array
$nin	Will NOT match the target value with ANY of the test values in the array

Table 5.2 – MongoDB comparison operators

Update operators

The following is a list of all update operators that MongoDB supports:

Name	Description
$inc	Will increment the value of the target field by the specified value.
$mul	Will multiply the value of the target field by the specified value.
$rename	Will rename a field.
$setOnInsert	Will set the value of the target field if an update results in an insert of a document. Will not modify an existing document or apply to any update operations.
$set	Will set the value of the target field to our specified value.
$unset	Will remove the target field from the document.
$min	Will update the target field to our specified value if the target value is less than the existing one.
$max	Will update the target field to our specified value if the target value is greater than the existing one.
$currentDate	Will set the value of the target field to the present date. We can define the value as a timestamp (Unix epoch time) or a date.

Table 5.3 – MongoDB update operators

Smart querying

There are several considerations that we have to take into account when querying in MongoDB. Let's look at some best practices for using regular expressions, query results, and cursors, and when deleting documents.

Using regular expressions

MongoDB offers a rich interface for querying using regular expressions. In its simplest form, we can use regular expressions in queries by modifying the query string:

```
> db.books.find({"name": /mongo/})
```

This is done to search for books in our books collection that contain the mongo name. It is the equivalent of a SQL LIKE query.

> **Note/Tip/Important Note**
> MongoDB uses **Perl Compatible Regular Expression (PCRE)** version 8.42 with UTF-8 support.

We can also use some options when querying:

Option	Description
i	This option queries case insensitivity.
m	This option only applies for multiline strings with anchors ($\wedge$ for the start and $ for the end). In this case, defining the m option will match the pattern at the beginning or end of each line. Without the m option, the anchors will match at the beginning or end of the string.

Table 5.4 – Querying options

In our previous example, if we wanted to search for mongo, Mongo, MONGO, or any other case-insensitive variation, we would need to use the i option, as follows:

```
> db.books.find({"name": /mongo/i})
```

Alternatively, we can use the $regex operator, which provides more flexibility.

The same queries using $regex will be written as follows:

```
> db.books.find({'name': { '$regex': /mongo/ } })
> db.books.find({'name': { '$regex': /mongo/i } })
```

By using the $regex operator, we can also use the options shown in the following table:

Option	Description
x	This option will ignore all whitespace characters in the $regex pattern. It will also ignore any content between and including an unescaped hash or pound character and the next newline. This can be used to include comments. It will not have any effect if the characters are included in a character class or escaped otherwise. It will not have any effect on handling the VT character.
s	This option allows the dot character (that is, .) to match all characters, including newline characters.

Table 5.5 – Regular expressions – additional operators

Expanding matching documents using regex makes our queries slower to execute.

Indexes that use regular expressions can only be used if our regular expression does queries for the beginning of a string that is indexed; that is, regular expressions starting with $\wedge$ or \A. If we only want to query using a starts with regular expression, we should avoid writing lengthier regular expressions, even if they will match the same strings.

Take the following code block as an example:

```
> db.books.find({'name': { '$regex': /mongo/ } })
> db.books.find({'name': { '$regex': /^mongo.*/ } })
```

Both queries will match name values starting with mongo (case-sensitive), but the first one will be faster as it will stop matching as soon as it hits the sixth character in every name value.

Querying results and cursors

MongoDB does not follow relational database transaction semantics by default. Instead, MongoDB queries will not belong to a transaction unless we explicitly define a transaction wrapper, as explained in *Chapter 6, Multi-Document ACID Transactions*.

Updates can modify the size of a document. Modifying the size can result in MongoDB moving the document on the disk to a new slot toward the end of the storage file.

When we have multiple threads querying and updating a single collection, we can end up with a document appearing multiple times in the result set.

This will happen in the following scenarios:

- Thread A starts querying the collection and matches the A1 document.

- Thread B updates the A1 document, increasing its size and forcing MongoDB to move it to a different physical location toward the end of the storage file.

- Thread A is still querying the collection. It reaches the end of the collection and finds the A1 document again with its new value, as shown here:

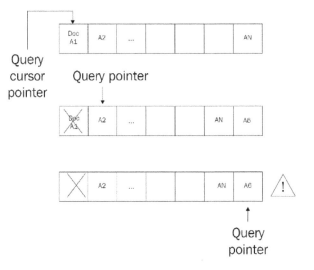

Figure 5.1 – Cursors and querying

This is rare, but it can happen in production; if we can't safeguard from such a case in the application layer, we can use hint ({ $natural : 1 }) to prevent it.

hint ({ $natural : 1 }) is supported by official drivers and the shell by appending it into an operation that returns a cursor:

```
> db.books.find().hint( { $natural : 1 } )
```

The older snapshot () option has been removed and using $natural can return the results in ascending or descending order (by substituting 1 with −1 in the order direction).

If our query runs on a unique index of a field whose values won't be modified during the duration of the query, we should use this query to get the same query behavior. Even then, $natural cannot protect us from insertions or deletions happening in the middle of a query. The $natural operator will traverse the built-in index that every collection has on the id field, making it inherently slow. This should only be used as a last resort.

If we want to update, insert, or delete multiple documents without other threads seeing the results of our operation while it's happening, we should use MongoDB transactions, as explained in the next chapter.

Storage considerations for the delete operation

Deleting documents in MongoDB does not reclaim the disk space used by it. If we have 10 GB of disk space used by MongoDB and we delete all documents, we will still be using 10 GB. Under the hood, MongoDB marks these documents as deleted and may use the space to store new documents.

This results in our disk having space that is not used but is not freed up for the operating system. If we want to claim it back, we can use compact ():

```
> db.books.compact()
```

Alternatively, we can start the mongod server with the --repair option.

Another option is to enable compression, which is available from version 3.0 and only with the WiredTiger storage engine. We can use the snappy or zlib algorithms to compress our document's size. This will, again, not prevent storage holes, but if we are tight on disk space, it is preferable to the heavy operational route of repair and compact.

Storage compression uses less disk space at the expense of CPU usage, but this trade-off is mostly worth it.

> **Note**
>
> Always take a backup before running operations that can result in a catastrophic loss of data. Repair or compact will run in a single thread, blocking the entire database from other operations. In production systems, always perform these on the secondary server first; then, switch the primary-secondary roles and compact the ex-primary, now-secondary instance.

In this section, we learned all the different ways we can query, update, insert, and delete documents in MongoDB using a range of languages with direct operations using the drivers or using ODM. In the next section, we will learn about change streams, a feature that allows our application to react immediately to changes in the underlying data.

Change streams

The change streams functionality was introduced in version 3.6 and updated in versions 4.0 and 5.1, making it a safe and efficient way to listen for database changes.

Introduction

The fundamental problem that change streams solve is the need for applications to react immediately to changes in the underlying data. Modern web applications need to be reactive to data changes and refresh the page view without reloading the entire page. This is one of the problems that frontend frameworks (such as Angular, React, and Vue.js) are solving. When a user performs an action, the frontend framework will submit the request to the server asynchronously and refresh the relevant fragment of the page based on the response from the server.

Thinking of a multiuser web application, there are cases where a database change may have occurred as a result of another user's action. For example, in a project management Kanban board, user A may be viewing the Kanban board, while another user, B, may be changing the status of a ticket from "To do" to "In progress."

User A's view needs to be updated with the change that user B has performed in real time, without refreshing the page. There are already three approaches to this problem, as follows:

- The most simple approach is to poll the database every X number of seconds and determine if there has been a change. Usually, this code will need to use some kind of status, timestamp, or version number to avoid fetching the same change multiple times. This is simple, yet inefficient, as it cannot scale with a great number of users. Having thousands of users polling the database at the same time will result in a high database-locking rate.

- To overcome the problems imposed by the first approach, database-and application-level triggers have been implemented. A database trigger relies on the underlying database executing some code in response to a database change. However, the main downside is, again, similar to the first approach in that the more triggers that we add to a database, the slower our database will become. It is also coupled to the database, instead of being a part of the application code base.

- Finally, we can use the database transaction or replication log to query for the latest changes and react to them. This is the most efficient and scalable approach of the three as it doesn't put a strain on the database. The database writes to this log anyway; it is usually appended only and our background task serially reads entries as they come into the log. The downside of this method is that it is the most complicated one to implement and one that can lead to nasty bugs if it's not implemented properly.

Change streams provide a way to solve this problem that is developer-friendly and easy to implement and maintain. Change streams are based on the oplog, which is MongoDB's operations log and contains every operation happening server-wide across all databases on the server. This way, the developer does not have to deal with the server-wide oplog or tailable cursors, which are often not exposed or as easy to develop from the MongoDB language-specific drivers. Also, the developer does not have to decipher and understand any of the internal oplog data structures that are designed and built for MongoDB's benefit, and not for an application developer.

Change streams also have other advantages around security:

- Users can only create change streams on collections, databases, or deployments that they have read access to.

- Change streams are also idempotent by design. Even in the case that the application cannot fetch the absolute latest change stream event notification ID, it can resume applying from an earlier known one and it will eventually reach the same state.

- Finally, change streams are resumable. Every change stream response document includes a resume token. If the application gets out of sync with the database, it can send the latest resume token back to the database and continue processing from there. This token needs to be persisted in the application, as the MongoDB driver won't keep application failures and restarts. It will only keep state and retry in case of transient network failures and MongoDB replica set elections.

Setup

A change stream can be opened against a collection, a database, or an entire deployment (such as a replica set or sharded cluster). A change stream will not react to changes in any system collection or any collection in the admin, config, and local databases.

A change stream requires a WiredTiger storage engine and replica set protocol version 1 (pv1). pv1 is the only supported version starting from MongoDB 4.0. Change streams are compatible with deployments that use encryption-at-rest.

Using change streams

To use a change stream, we need to connect to our replica set. A replica set is a prerequisite to using change streams. As change streams internally use the oplog, it's not possible to work without it. Change streams will also output documents that won't be rolled back in a replica set setting, so they need to follow a majority read concern. Either way, it's a good practice to develop and test locally using a replica set, as this is the recommended deployment for production. As an example, we are going to use a `signals` collection within our database named `streams`.

We will use the following sample Python code:

```python
from pymongo import MongoClient


class MongoExamples:
    def __init__(self):
        self.client = MongoClient('localhost', 27017)
        db = self.client.streams
        self.signals = db.signals
    # a basic watch on signals collection
    def change_books(self):
        with self.client.watch() as stream:
            for change in stream:
                print(change)
def main():
    MongoExamples().change_books()
if __name__ == '__main__':
    main()
```

We can open one Terminal and run it using `python change_streams.py`.

Then, in another Terminal, we connect to our MongoDB replica set using the following code:

```
> mongo
> use streams
> db.signals.insert({value: 114.3, signal:1})
```

Going back to our first Terminal window, we can now observe that the output is similar to the following code block:

```
{'_id': {'_data': '825BB7A25E0000000129295A1004A34408FB-
07864F8F960BF14453DFB98546645F696400645BB7A25EE10ED33145B-
CF7A70004'}, 'operationType': 'insert', 'clusterTime':
Timestamp(1538761310, 1), 'fullDocument': {'_id': Objec-
tId('5bb7a25ee10ed33145bcf7a7'), 'value': 114.3, 'signal':
1.0}, 'ns': {'db': 'streams', 'coll': 'signals'}, 'document-
Key': {'_id': ObjectId('5bb7a25ee10ed33145bcf7a7')}}
```

Here, we have opened a cursor that's watching the entire `streams` database for changes. Every data update in our database will be logged and outputted in the console.

For example, if we go back to the mongo shell, we can issue the following code:

```
> db.a_random_collection.insert({test: 'bar'})
```

The Python code output should be similar to the following code:

```
{'_id': {'_data': '825BB7A3770000000229295A10044AB37F707D-
104634B646CC5810A40EF246645F696400645BB7A377E10ED33145B-
CF7A80004'}, 'operationType': 'insert', 'clusterTime':
Timestamp(1538761591, 2), 'fullDocument': {'_id': Objec-
tId('5bb7a377e10ed33145bcf7a8'), 'test': 'bar'}, 'ns': {'db':
'streams', 'coll': 'a_random_collection'}, 'documentKey': {'_
id': ObjectId('5bb7a377e10ed33145bcf7a8')}}
```

This means that we are getting notifications for every data update across all the collections in our database.

Now, we can change line 11 of our code to the following:

```
> with self.signals.watch() as stream:
```

This will result in only watching the `signals` collection, as should be the most common use case.

PyMongo's `watch` command can take several parameters, as follows:

```
watch(pipeline=None, full_document='default', resume_
after=None, max_await_time_ms=None, batch_size=None,
collation=None, start_at_operation_time=None, session=None)
```

The most important parameters are as follows:

- `Pipeline`: This is an optional parameter that we can use to define an aggregation pipeline to be executed on each document that matches `watch()`. Because the change stream itself uses the aggregation pipeline, we can attach events to it. The aggregation pipeline events we can use are as follows:

  ```
  $match
  $project
  $addFields
  $replaceRoot
  $redact
  $replaceWith
  ```

```
$set
$unset
```

- **Full_document**: This is an optional parameter that we can use by setting it to `'updateLookup'` to force the change stream to return a copy of the document as it has been modified in the `fullDocument` field, along with the document's delta in the `updateDescription` field. The default None value will only return the document's delta.

- **start_at_operation_time**: This is an optional parameter that we can use to only watch for changes that occurred at, or after, the specified timestamp.

- **session**: This is an optional parameter in case our driver supports passing a `ClientSession` object to watch for updates.

Change streams response documents have to be under 16 MB in size. This is a global limit in MongoDB for BSON documents and the change stream has to follow this rule.

Specification

The following document shows all of the possible fields that a change event response may or may not include, depending on the actual change that happened:

```
{ _id : { <BSON Object> },
  "operationType" : "<operation>",
  "fullDocument" : { <document> },
  "ns" : {
     "db" : "<database>",
     "coll" : "<collection"
  },
  "documentKey" : { "_id" : <ObjectId> },
  "updateDescription" : {
     "updatedFields" : { <document> },
     "removedFields" : [ "<field>", ... ]
  }
  "clusterTime" : <Timestamp>,
  "txnNumber" : <NumberLong>,
  "lsid" : {
     "id" : <UUID>,
     "uid" : <BinData>
  }
}
```

The most important fields are as follows:

`fullDocument`	This is the new state of the document, which can include the following: If it's a delete operation, this field is omitted as the document no longer exists. If it's an insert or replace operation, this will be the new value of the document. If it's an update operation and we have enabled `'update-Lookup'`, then it will have the most recently major-committed version of the document modified by the update operation.
`operationType`	This is the type of operation; it can be either `insert`, `delete`, `replace`, `update`, or `invalidate`.
`documentKey`	This is the ID (`ObjectID`) of the document that was modified by the operation.
`updateDescription.updatedFields` / `removedFields`	This is a document or an array of keys showing the data that was updated or removed by the update or remove operation, respectively.
`txnNumber`	The transaction ID number when the operation is part of a multi-document atomicity, consistency, isolation, durability (ACID) transaction.
`lsid`	This is the session ID when the operation is part of a multi-document ACID transaction.

Table 5.6 – Change streams – the most common events

Important notes

When using a sharded database, change streams need to be opened against a MongoDB server. When using replica sets, a change stream can only be opened against any data-bearing instance. Each change stream will open a new connection, as of 4.0.2. If we want to have lots of change streams in parallel, we need to increase the connection pool (as per the SERVER-32946 JIRA MongoDB ticket) to avoid severe performance degradation.

Production recommendations

Let's look at some of the best recommendations by MongoDB and expert architects at the time of writing.

Replica sets

Starting from MongoDB 4.2, a change stream can still be available even if the Read Concern of the majority is not satisfied. The way to enable this behavior is by setting { `majority : false` }.

Invalidating events, such as dropping or renaming a collection, will close the change stream. We cannot resume a change stream after an invalidating event closes it.

As the change stream relies on the oplog size, we need to make sure that the oplog size is large enough to hold events until they are processed by the application.

We can open a change stream operation against any data-bearing member in a replica set.

Sharded clusters

On top of the considerations for replica sets, there are a few more to keep in mind for sharded clusters. They are as follows:

- The change stream is executed against every shard in a cluster and will be as fast as the slowest shard
- To avoid creating change stream events for orphaned documents, we need to use the new feature of ACID-compliant transactions if we have multi-document updates under sharding

We can only open a change stream operation against the mongos member in a sharded cluster.

While sharding an unsharded collection (that is, migrating from a replica set to sharding), the documentKey property of the change stream notification document will include _id until the change stream catches up to the first chunk migration.

Queryable encryption

MongoDB 6 introduced end-to-end queryable encryption. Seny Kamara and Tarik Moataz, from Brown University, designed *An Optimal Relational Database Encryption Scheme* at Aroki Systems that was acquired by MongoDB in 2021.

Queryable encryption allows database queries, while transmitted, to be processed and stored data to ensure they are always encrypted. The developer uses the MongoDB driver the same way as always, but the driver handles encrypting the query and the database queries the data without ever decrypting the query or the data.

Queryable encryption allows developers to query encrypted data fields as easily as they would do with the non-encrypted fields without having to learn cryptography. It provides the fundamental controls to meet data privacy regulatory requirements such as GDPR, HIPAA, CCPA, and other related legislation. It can reduce operational risk for migrating on-premise datasets to the cloud by providing the peace of mind that data will be encrypted end to end, from querying and at rest, to in-transit data processing.

The closest equivalent to queryable encryption is homomorphic encryption, which also allows computations to be performed on encrypted data without the need to decrypt them. One of the most advanced libraries in the field is Microsoft SEAL (https://github.com/microsoft/SEAL/) which, as of Summer 2022, is far more complex and less feature complete than MongoDB's offering. Other notable implementations are IBM security's **Fully Homomorphic Encryption (FHE)** and the SDK developed by zama.ai.

The way that queryable encryption is implemented in MongoDB is that the driver needs to connect to a customer-provisioned key provider. This can be a cloud provider Key Management System (for example, AWS KMS or Azure Key Vault), or an on-premises HSM/key service provider. From that point of view, the client-server communication, querying, and data transmission are completely transparent to the developer.

MongoDB Atlas and the Enterprise version support automatic encryption, while the Community Edition requires writing application code to encrypt and decrypt the data fields at the time of writing.

At the time of writing, MongoDB drivers support the following methods with queryable encryption:

- `aggregate`
- `count`
- `delete`
- `distinct`
- `explain`
- `find`
- `findAndModify`
- `insert`
- `update`

These methods support the following operators:

- `$eq`
- `$ne`
- `$in`
- `$nin`
- `$and`
- `$or`
- `$not`
- `$nor`
- `$set`
- `$unset`

Aggregation supports the following stages:

- `$addFields`
- `$bucket`
- `$bucketAuto`
- `$collStats`
- `$count`
- `$geoNear`
- `$indexStats`
- `$limit`
- `$match`
- `$project`
- `$redact`
- `$replaceRoot`
- `$sample`
- `$skip`
- `$sort`
- `$sortByCount`
- `$unwind`
- `$group (with limitations)`
- `$lookup (with limitations)`
- `$graphLookup (with limitations)`

These stages support the following aggregation expressions:

- `$cond`
- `$eq`
- `$ifNull`
- `$in`
- `$let`
- `$literal`

- `$ne`

- `$switch`

Support will greatly improve in subsequent versions. The MongoDB official documentation at `https://www.mongodb.com/docs/manual/core/queryable-encryption/` is the best place to keep up to date with the latest progress.

Summary

In this chapter, we went through advanced querying concepts while using Ruby, Python, and PHP alongside the official drivers and an ODM.

Using Ruby and the Mongoid ODM, Python and the PyMODM ODM, and PHP and the Doctrine ODM, we went through code samples to explore how to `create`, `read`, `update`, and `delete` documents.

We also discussed batching operations for performance and best practices. We presented an exhaustive list of comparison and update operators that MongoDB uses.

Following that, we discussed smart querying, how cursors in querying work, what our storage performance considerations should be on delete, and how to use regular expressions.

Next, we discussed change streams and how we can use them to subscribe and receive updates in our database in real time.

Finally, we learned about one of the major features that was introduced in MongoDB 6: queryable encryption.

In the next chapter, we will learn about the aggregation framework by covering a complete use case that involves processing transaction data from the Ethereum blockchain.

6

Multi-Document ACID Transactions

MongoDB introduced multi-document ACID transactions (distributed transactions) in July 2018 with the release of v4.0.

Relational database transactions follow the **atomicity**, **consistency**, **isolation**, and **durability** (ACID) database principles, through the use of transactions. MongoDB, as a document-oriented database, doesn't need ACID transactions in general. There are a few cases, though, where ACID properties are needed and that's when we should, as an exception, use multi-document ACID transactions in MongoDB.

We will explore multi-document ACID transactions (also known as distributed transactions) in MongoDB and learn where they can be a useful alternative to the CRUD operators that we learned about in the previous chapters. We will use two cases – building a digital bank and developing an e-commerce cart – to understand how distributed transactions can help us achieve consistency and correct implementation using these semantics.

In the previous chapter, we learned how to query MongoDB using Ruby, Python, and PHP drivers and frameworks.

In this chapter, we will cover the following topics:

- Transactions background
- Exploring ACID properties
- E-commerce using MongoDB

Technical requirements

To follow along with the code in this chapter, you need to install MongoDB locally, set up a replica set, and install the Ruby and Python language drivers.

Transactions background

MongoDB is a non-relational database and provides only a few guarantees around ACID. Data modeling in MongoDB does not focus on BCNF, 2NF, and 3NF normalization; instead, its focus is in the opposite direction.

In MongoDB, the best approach is often to embed our data into subdocuments, resulting in more self-contained documents than a single row of data in an RDBMS. This means that a logical transaction can affect a single document many times. Single-document transactions are ACID-compliant in MongoDB, meaning that multi-document ACID transactions have not been essential for MongoDB development.

However, there are a few reasons why getting multi-document transactions is a good idea. Over the years, MongoDB has grown from being a niche database to a multi-purpose database that is used everywhere – from start-ups to major Fortune 500 companies. Across many different use cases, there are inevitably a few corner cases where data modeling can't, or shouldn't, fit data in subdocuments and arrays.

A great example is an unbounded one-to-many relationship. When we expect an unbounded number of documents in the "many" part of the relationship, the document can grow past the maximum document object size of 16 MB. In this case, we need a mechanism to create a chain of documents that will all stay under the maximum document object size.

Also, even when the best solution for a data architect today is to embed data, they can't be sure this will always be the case. This makes choosing the right database layer difficult.

RDBMS data modeling has been around for over 40 years and is a well-known and understood data modeling process. Helping data architects work with something they are familiar with is always a bonus.

Before multi-document transactions were introduced, the only workaround was implementing them in a customized way in the application layer. This was both time-consuming and error-prone. Implementing a two-phase commit process in the application layer could also be slower and lead to increased locking, slowing our database operations.

In this chapter, we will focus on using the native MongoDB transactions, which are now strongly recommended by MongoDB Inc. In the next section, we will explore ACID, and what each of the elements means for us.

Exploring ACID properties

ACID stands for atomicity, consistency, isolation, and durability. In the following sections, we will explain what each of these means for our database design and architecture.

Atomicity

Atomicity refers to the concept that a transaction needs to follow the binary success or fail principle. If a transaction succeeds, then its results are visible to every subsequent user. If a transaction fails, then every change is rolled back to the point it was right before it started. Either all actions in a transaction occur or none at all.

A simple example to understand atomicity is by transferring money from account *A* to account *B*. Money needs to be credited from account *A* and then debited into account *B*. If the operation fails midway, then both accounts *A* and *B* need to be reverted to their state before the operation started.

In MongoDB, operations in a single document are always atomic even if the operation spans multiple subdocuments or arrays within the document.

Operations spanning multiple documents need to use MongoDB transactions to be made atomic.

Consistency

Consistency refers to the database's state. Every database operation must start and regardless of whether it succeeds or fails, it should leave the database in a state where its data is consistent. The database must always be in a consistent state.

Database constraints must be respected at all times. Any future transaction must also be able to view data that's been updated by past transactions. The consistency model most commonly used in practice for distributed data systems is eventual consistency.

Eventual consistency guarantees that once we stop updating our data, all future reads will eventually read the latest committed write value. In distributed systems, this is the only acceptable model in terms of performance, as data needs to be replicated over the network across different servers.

In contrast, the least popular model of strong consistency guarantees that every future read will always read the write value that was committed last. This implies that every update is propagated and committed to every server before the next read comes in, which will cause a huge strain on performance for these systems.

MongoDB falls somewhere in between eventual and strict consistency by adopting a causal consistency model. With causal consistency, any transaction execution sequence is the same as if all causally related read/write operations were executed in an order that reflects their causality.

What this means in practice is that concurrent operations may be seen in different orders and reads correspond to the latest value written about the writes that they are causally dependent on.

Eventually, it's a trade-off between how many concurrent operations can happen at once and the consistency of data being read by the application.

Isolation

Database isolation refers to the view that each transaction has of other transactions that run in parallel. Isolation protects us from transactions acting on the state of parallel running, incomplete transactions that may subsequently roll back. An example of why isolation levels are essential is described in the following scenario:

1. Transaction *A* updates user 1's account balance from £50 to £100 but does not commit the transaction.
2. Transaction *B* reads user 1's account balance as £100.
3. Transaction *A* is rolled back, reverting user 1's account balance to £50.
4. Transaction *B* thinks that user 1 has £100, whereas they only have £50.
5. Transaction *B* updates user 2's value by adding £100. User 2 receives £100 out of thin air from user 1, since user 1 only has £50 in their account. Our imaginary bank is in trouble.

Isolation typically has four levels, as follows, listed from the most to the least strict:

- Serializable
- Repeatable read
- Read committed
- Read uncommitted

The problems we can run into, from the least to the most serious, depend on the isolation level, as follows:

- Phantom reads
- Non-repeatable reads
- Dirty reads
- Lost updates

Losing data about an operational update is the worst thing that can happen in any database because this would render our database unusable and make it a store of data that cannot be trusted. That's why, in every isolation level, even read uncommitted isolation will not lose data.

However, the other three issues may also arise. We will briefly explain what these are in the following sections.

Phantom reads

A phantom read occurs when, during a transaction, another transaction modifies its result set by adding or deleting rows that belong to its result set. An example of this is as follows:

1. Transaction *A* queries for all users. 1,000 users are returned but the transaction does not commit.

2. Transaction *B* adds another user; 1,001 users are now in our database.

3. Transaction *A* queries for all users for a second time. 1,001 users are now returned. Transaction *A* now commits.

Under a strict serializable isolation level, transaction *B* should be blocked from adding the new user until transaction *A* commits its transaction. This can, of course, cause huge contention in the database, which can lead to performance degradation. This is because every update operation needs to wait for reads to commit their transactions. This is why, typically, the serializable level is rarely used in practice.

Non-repeatable reads

A non-repeatable read occurs when, during a transaction, a row is retrieved twice, and the row's values are different with every read operation.

Following the previous money transfer example, we can illustrate a non-repeatable read similarly:

1. Transaction *B* reads user 1's account balance as £50.

2. Transaction *A* updates user 1's account balance from £50 to £100 and commits the transaction.

3. Transaction *B* reads user 1's account balance again and gets the new value, £100, and then commits the transaction.

The problem here is that transaction *B* has got a different value in the course of its transaction because it was affected by transaction *A*'s update. This is a problem because transaction *B* is getting different values within its own transaction. However, in practice, it solves the issue of transferring money between users when they don't exist.

This is why a read committed isolation level, which does not prevent non-repeatable reads but does prevent dirty reads, is the most commonly used isolation level in practice.

Dirty reads

The previous example, where we made money out of thin air and ended up transferring £100 out of an account that only had £50 in balance, is a classic example of a dirty read.

A read uncommitted isolation level does not protect us from dirty reads and that is why it is rarely used in production-level systems.

The following are the isolation levels versus their potential issues:

Isolation Level	Lost Updates	Dirty Reads	Non-Repeatable Reads	Phantoms
Read uncommitted	Don't occur	May occur	May occur	May occur
Read committed	Don't occur	Don't occur	May occur	May occur
Repeatable read	Don't occur	Don't occur	Don't occur	May occur
Serializable	Don't occur	Don't occur	Don't occur	Don't occur

Table 6.1 – Database isolation levels and potential issues

PostgreSQL uses a default (and configurable) isolation level of read committed. As MongoDB is not inherently an RDBMS, using transactions for every operation makes the situation more complicated.

The equivalent isolation level in these terms is read uncommitted. This may look scary based on the examples given previously, but on the other hand, in MongoDB, there is (again, in general) no concept of transactions or rolling them back. Read uncommitted refers to the fact that changes will be made visible before they are made durable. More details on the made durable part will be provided in the next section.

Durability

Durability is all about data resilience in the face of failure. Every transaction that has successfully committed its state must survive failure. This is usually done by persisting the transaction results in persistent storage (SSD/HDD).

RDBMSs always follow the durability concept by writing every committed transaction to a transaction log or **write-ahead log** (**WAL**). MongoDB, using the WiredTiger storage engine, commits writes using WAL to its persistent storage-based journal every 60 milliseconds and is, for all practical purposes, durable. As durability is important, every database system prefers relaxing other aspects of ACID first, and durability usually gets relaxed last.

When do we need ACID in MongoDB?

Existing atomicity guarantees, for single-document operations, that MongoDB can meet the integrity needs of most real-world applications. However, some use cases have traditionally benefited from ACID transactions; modeling them in MongoDB could be significantly more difficult than using the well-known ACID paradigm.

Unsurprisingly, many of these cases come from the financial industry. Dealing with money and stringent regulation frameworks means that every operation needs to be stored, sometimes in a strict execution order so that they can be logged, verified, and audited if requested. Building a digital bank requires interaction between multiple accounts that could be represented as documents in MongoDB.

Managing high volumes of financial transactions, either by users or algorithms executing high-frequency trading, also requires verifying every single one of them. These transactions may span multiple documents as they would, again, refer to multiple accounts.

The general pattern for using multi-document ACID transactions is when we can have an unbounded number of entities, sometimes in the millions. In this case, modeling entities in subdocuments and arrays cannot work because the document would eventually outgrow the built-in 16 MB document size limit present in MongoDB.

Building a digital bank using MongoDB

The most common use cases for multi-document ACID transactions come from the financial sector. In this section, we will model a digital bank using transactions and go through progressively more complicated examples of how we can use transactions for our benefit.

The basic functionality that a bank must provide is accounts and transferring monetary amounts between them. Before transactions were introduced, MongoDB developers had two options. The first option – the MongoDB way of doing it – is to embed data in a document, either as a subdocument or as an array of values. In the case of accounts, this could result in a data structure such as the following:

```
{accounts: [ {account_id: 1, account_name: 'alex', balance:
100}, {account_id: 2, account_name: 'bob', balance: 50}]}
```

However, even in this simple format, it will quickly outgrow the fixed 16 MB document limit in MongoDB. The advantage of this approach is that since we have to deal with a single document, all operations will be atomic, resulting in strong consistency guarantees when we transfer money from one account to another.

The only viable alternative, except for using a relational database, is to implement guarantees at the application level that will simulate a transaction with the appropriate code in place to undo parts, or the whole, of a transaction in case of an error. This can work, but will result in a longer time to market and is more error-prone.

MongoDB's multi-document ACID transactions approach is similar to how we would work with transactions in a relational database. Taking the most simple example from MongoDB Inc.'s white paper, *MongoDB Multi-Document ACID Transactions*, published in June 2018, the generic transaction in MongoDB will look as follows:

```
s.start_transaction()
orders.insert_one(order, session=s)
stock.update_one(item, stockUpdate, session=s)
s.commit_transaction()
```

However, the same transaction in MySQL will look as follows:

```
db.start_transaction()
cursor.execute(orderInsert, orderData)
cursor.execute(stockUpdate, stockData)
db.commit()
```

That said, in modern web application frameworks, most of the time, transactions are hidden in the **object-relational mapping (ORM)** layer and not immediately visible to the application developer. The framework ensures that web requests are wrapped in transactions to the underlying database layer. This is not yet the case for ODM frameworks, but you would expect that this could now change.

Setting up our data

We are going to use a sample init_data.json file with two accounts. Alex has 100 of the hypnotons imaginary currency, whereas Mary has 50 of them:

```
{"collection": "accounts", "account_id": "1", "account_name":
"Alex", "account_balance":100}{"collection": "accounts",
"account_id": "2", "account_name": "Mary", "account_
balance":50}
```

Using the following Python code, we can insert these values into our database:

```
import json
class InitData:
def __init__(self):
self.client = MongoClient('localhost', 27017)
self.db = self.client.mongo_bank
self.accounts = self.db.accounts
# drop data from accounts collection every time to start from a
clean slate
self.accounts.drop()
# load data from json and insert them into our database
init_data = InitData.load_data(self)
self.insert_data(init_data)
@staticmethod
def load_data(self):
ret = []
with open('init_data.json', 'r') as f:
for line in f:
```

```
ret.append(json.loads(line))
return ret
def insert_data(self, data):
for document in data:
collection_name = document['collection']
account_id = document['account_id']
account_name = document['account_name']
account_balance = document['account_balance']
self.db[collection_name].insert_one({'account_id': account_id,
'name': account_name, 'balance': account_balance})
```

This results in our `mongo_bank` database having the following documents in the `accounts` collection:

```
> db.accounts.find()
{ "_id" : ObjectId("5bc1fa7ef8d89f2209d4afac"), "account_id" :
"1", "name" : "Alex", "balance" : 100 }
{ "_id" : ObjectId("5bc1fa7ef8d89f2209d4afad"), "account_id" :
"2", "name" : "Mary", "balance" : 50 }
```

Transferring between accounts – part 1

As a MongoDB developer, the most familiar way to model a transaction is to implement basic checks in the code. With our sample account documents, you may be tempted to implement an account transfer as follows:

```
def transfer(self, source_account, target_account, value):
print(f'transferring {value} Hypnotons from {source_account} to
{target_account}')
with self.client.start_session() as ses:
ses.start_transaction()
self.accounts.update_one({'account_id': source_account},
{'$inc': {'balance': value*(-1)} })
self.accounts.update_one({'account_id': target_account},
{'$inc': {'balance': value} })
updated_source_balance = self.accounts.find_one({'account_id':
source_account})['balance']
updated_target_balance = self.accounts.find_one({'account_id':
target_account})['balance']
if updated_source_balance < 0 or updated_target_balance < 0:
```

```
ses.abort_transaction()
else:
ses.commit_transaction()
```

Calling this method in Python will transfer 300 hypnotons from account 1 to account 2:

```
>>> obj = InitData.new
>>> obj.transfer('1', '2', 300)
```

This will result in the following output:

```
> db.accounts.find()
{ "_id" : ObjectId("5bc1fe25f8d89f2337ae40cf"), "account_id" :
"1", "name" : "Alex", "balance" : -200 }
{ "_id" : ObjectId("5bc1fe26f8d89f2337ae40d0"), "account_id" :
"2", "name" : "Mary", "balance" : 350 }
```

The problem here isn't the checks on `updated_source_balance` and `updated_target_balance`. Both of these values reflect the new values of -200 and 350, respectively. The problem isn't the `abort_transaction()` operation either. Instead, the problem is that we are not using the session.

The single most important thing to learn about transactions in MongoDB is that we need to use the session object to wrap operations in a transaction. However, we can still perform operations outside the transaction scope within a transaction code block.

What happened here is that we initiated a transaction session, as follows:

```
with self.client.start_session() as ses:
```

Then, we completely ignored it by doing all of our updates in a non-transactional way. After that, we invoked `abort_transaction`, as follows:

```
ses.abort_transaction()
```

The transaction to be aborted was essentially void and didn't have anything to roll back.

Transferring between accounts – part 2

The correct way to implement a transaction is to use the session object in every operation that we want to either commit or roll back at the end of it, as shown in the following code:

```
def tx_transfer_err(self, source_account, target_account,
value):
```

```
print(f'transferring {value} Hypnotons from {source_account} to
{target_account}')
with self.client.start_session() as ses:
ses.start_transaction()
res = self.accounts.update_one({'account_id': source_account},
{'$inc': {'balance': value*(-1)} }, session=ses)
res2 = self.accounts.update_one({'account_id': target_account},
{'$inc': {'balance': value} }, session=ses)
error_tx = self.__validate_transfer(source_account, target_
account)

if error_tx['status'] == True:
print(f"cant transfer {value} Hypnotons from {source_account}
({error_tx['s_bal']}) to {target_account} ({error_tx['t_
bal']})")
ses.abort_transaction()
else:
ses.commit_transaction()
```

The only difference now is that we are passing `session=ses` in both of our update statements. To validate whether we have enough funds to make the transfer, we wrote a helper method, `__validate_transfer`, with its arguments being the source and target account IDs:

```
def __validate_transfer(self, source_account, target_account):
source_balance = self.accounts.find_one({'account_id': source_
account})['balance']
target_balance = self.accounts.find_one({'account_id': target_
account})['balance']

if source_balance < 0 or target_balance < 0:
return {'status': True, 's_bal': source_balance, 't_bal':
target_balance}
else:
return {'status': False}
```

Unfortunately, this attempt will also *fail*. The reason is the same as before. When we are inside a transaction, we make changes to the database that follow the ACID principles. Changes inside a transaction are not visible to any queries outside of it until they are committed.

Transferring between accounts – part 3

The correct implementation of the transfer problem will look as follows (the full code sample can be found in this book's GitHub repository):

```python
from pymongo import MongoClient
import json


class InitData:
def __init__(self):
self.client = MongoClient('localhost', 27017, w='majority')
self.db = self.client.mongo_bank
self.accounts = self.db.accounts


# drop data from accounts collection every time to start from a
clean slate
self.accounts.drop()

init_data = InitData.load_data(self)
self.insert_data(init_data)
self.transfer('1', '2', 300)

@staticmethod
def load_data(self):
ret = []
with open('init_data.json', 'r') as f:
for line in f:
ret.append(json.loads(line))
return ret

def insert_data(self, data):
for document in data:
collection_name = document['collection']
account_id = document['account_id']
account_name = document['account_name']
account_balance = document['account_balance']
```

```
self.db[collection_name].insert_one({'account_id': account_id,
'name': account_name, 'balance': account_balance})
```

Now, let's validate any errors using the following code:

```
# validating errors, using the tx session
def tx_transfer_err_ses(self, source_account, target_account,
value):
print(f'transferring {value} Hypnotons from {source_account} to
{target_account}')
with self.client.start_session() as ses:
ses.start_transaction()
res = self.accounts.update_one({'account_id': source_account},
{'$inc': {'balance': value * (-1)}}, session=ses)
res2 = self.accounts.update_one({'account_id': target_account},
{'$inc': {'balance': value}}, session=ses)
error_tx = self.__validate_transfer_ses(source_account, target_
account, ses)

if error_tx['status'] == True:
print(f"cant transfer {value} Hypnotons from {source_account}
({error_tx['s_bal']}) to {target_account} ({error_tx['t_
bal']})")
ses.abort_transaction()
else:
ses.commit_transaction()
```

Now, let's pass the session value so that we can view the updated values:

```
def __validate_transfer_ses(self, source_account, target_
account, ses):
source_balance = self.accounts.find_one({'account_id': source_
account}, session=ses)['balance']
target_balance = self.accounts.find_one({'account_id': target_
account}, session=ses)['balance']
if source_balance < 0 or target_balance < 0:
return {'status': True, 's_bal': source_balance, 't_bal':
target_balance}
else:
return {'status': False}
```

```
def main():
InitData()

if __name__ == '__main__':
main()
```

In this case, by passing the session object's `ses` value, we ensure that we can both make changes in our database using `update_one()` and also view these changes using `find_one()`, before doing either an `abort_transaction()` operation or a `commit_transaction()` operation.

In the end, if we need to roll back using transactions, we don't need to keep track of the previous account balance values. This is because MongoDB will discard all of the changes that we made inside the transaction scope.

Continuing with the same example using Ruby, we have the following code for part 3:

```
require 'mongo'

class MongoBank
def initialize
@client = Mongo::Client.new([ '127.0.0.1:27017' ], database:
:mongo_bank)
db = @client.database
@collection = db[:accounts]

# drop any existing data
@collection.drop

@collection.insert_one('collection': 'accounts', 'account_id':
'1', 'account_name': 'Alex', 'account_balance':100)
@collection.insert_one('collection': 'accounts', 'account_id':
'2', 'account_name': 'Mary', 'account_balance':50)

transfer('1', '2', 30)
transfer('1', '2', 300)
end

def transfer(source_account, target_account, value)
```

```ruby
puts "transferring #{value} Hypnotons from #{source_account} to
#{target_account}"
session = @client.start_session

session.start_transaction(read_concern: { level: :snapshot },
write_concern: { w: :majority })
@collection.update_one({ account_id: source_account }, { '$inc'
=> { account_balance: value*(-1)} }, session: session)
@collection.update_one({ account_id: target_account }, { '$inc'
=> { account_balance: value} }, session: session)

source_account_balance = @collection.find({ account_id: source_
account }, session: session).first['account_balance']

if source_account_balance < 0
session.abort_transaction
else
session.commit_transaction
end
end

end

# initialize class
MongoBank.new
```

Transactions can also perform some **data definition language** (**DDL**) operations, starting from version 4.4. A transaction can create a *collection or indexes in empty collections that have already been created in the same transaction.*

A collection can be created explicitly using the `create_collection()` operator or implicitly by creating or upserting a document targeting a collection that does not already exist.

There are still some limitations in place; for example, we cannot write/update across different shards in the same transaction. As an example, let's say our transaction is performing the following two operations:

- Adding or updating a document in the `accounts` collection in shard A
- Implicitly or explicitly creating a collection in shard B

Here, this transaction will abort.

Another limitation is that when we want to create a collection or index explicitly (that is, using the create collection and index methods), we need to set the transaction read concern to `local`.

Finally, we cannot list collections and indexes and we cannot use any other non-CRUD and non-informational operators. This would include methods such as `count()` and `createUser()`.

> **Note**
>
> We can still use the `count()` method to enumerate the number of documents inside a transaction by wrapping the command inside an aggregation framework operation and using `$count` or `$group` combined with `$sum`. MongoDB drivers will usually provide a helper method, `countDocuments(filter, options)`, that does exactly that.

MongoDB also allows us to customize `read_concern` and `write_concern` per transaction.

The available `read_concern` levels for multi-document ACID transactions are as follows:

- `majority`: The majority of the servers in a replica set have acknowledged the data. For this to work as expected in transactions, they must also use `write_concern` set to `majority`.

- `local`: Only the local server has acknowledged the data. This is the default `read_concern` level for transactions.

- `snapshot`: If the transaction commits with `majority` set to `write_concern`, all the transaction operations will have read from a snapshot of the majority of the committed data; otherwise, no guarantee can be made.

A snapshot read concern is also available outside of multi-document ACID transactions for the `find()` and `aggregate()` methods. It is also available for the `distinct()` method if the collection is not sharded.

> **Note**
>
> A read concern for transactions is set at the transaction level or higher (session or client). Setting a read concern in individual operations is not supported and is generally discouraged.

The available `write_concern` levels for multi-document ACID transactions are as follows:

- `majority`: The majority of the servers in a replica set have acknowledged the data. This is the **default** write concern as of MongoDB 5.0.

- `<w>`: The `<w>` number of servers have to acknowledge the write before it's considered successful. `w==1` will write to primary, `w==2` will write to primary and one data bearing node, and so on.

- `<custom_write_concern_name>`: We can also tag our servers and cluster them under `<custom_write_concern_name>`. This way, we can wait for acknowledgment from our desired number of nodes and also specify exactly which servers we want to propagate our writes to. This is useful, for example, when we have disaster recovery scenarios where one of the servers is hosted on another data center and we need to make sure that the writes are always propagated there. The operation may be rolled back if our `<custom_write_concern_name>` set only ends up having one server and that server steps down from being a primary before the write is acknowledged.

The default write concern will be different in the case that we have one or more arbiter nodes in our cluster. In this case, if the number of data nodes is less than or equal to the "majority number of voting nodes," then the write concern falls back to 1, instead of the default value of "majority."

The "majority number of voting nodes" is calculated as *1 + floor(<number_of_voting nodes>)*.

`floor()` will round down the number of voting nodes to the nearest integer.

The number of voting nodes is the sum of arbiter nodes plus the data-bearing nodes.

> **Note**
> Transaction read and write concerns will fall back from the transaction, then to the session level, and finally to the MongoDB client-level defaults if unset.

Some other transaction limitations as of MongoDB 5.3 are as follows:

- We can't write to any system or capped collections.
- We can't read or write any collections in the admin, config, or local databases.
- We can't inspect the query planner using `explain()`.
- We can't read from a capped collection using a snapshot read concern.
- The `getMore` operation on cursors must be created and accessed either inside or outside the transaction. This means that if we create the cursor inside a transaction we can only use `getMore()` on the cursor inside it. The same goes for creating it outside the transaction.
- We can't start a transaction by invoking `killCursors()` as the first operation.

In this section, we learned about ACID properties through the digital bank use case. In the next section, we will explore a more complex use case of using transactions for e-commerce.

E-commerce using MongoDB

For our second example, we are going to use a more complex use case on a transaction with three different collections.

We are going to simulate a shopping cart and payment transaction process for an e-commerce application using MongoDB. Using the sample code that we'll provide at the end of this section, we will initially populate the database with the following data.

Our first collection is the users collection with one document per user:

```
> db.users.find()
{ "_id" : ObjectId("5bc22f35f8d89f2b9e01d0fd"), "user_id" : 1,
"name" : "alex" }
{ "_id" : ObjectId("5bc22f35f8d89f2b9e01d0fe"), "user_id" : 2,
"name" : "barbara" }
```

Then, we have the carts collection with one document per cart, which is linked via user_id to our users:

```
> db.carts.find()
{ "_id" : ObjectId("5bc2f9de8e72b42f77a20ac8"), "cart_id" : 1,
"user_id" : 1 }
{ "_id" : ObjectId("5bc2f9de8e72b42f77a20ac9"), "cart_id" : 2,
"user_id" : 2 }
```

The payments collection holds any completed payment that has gone through, storing cart_id and item_id to link to the cart that it belonged to and the item that has been paid, respectively:

```
> db.payments.find()
{ "_id" : ObjectId("5bc2f9de8e72b42f77a20aca"), "cart_id" : 1,
"name" : "alex", "item_id" : 101, "status" : "paid" }
```

Finally, the inventories collection holds a count of the number of items (by item_id) that we currently have available, along with their price and a short description:

```
> db.inventories.find()
{ "_id" : ObjectId("5bc2f9de8e72b42f77a20acb"), "item_id" :
101, "description" : "bull bearing", "price" : 100, "quantity"
: 5 }
```

In this example, we are going to demonstrate using MongoDB's schema validation functionality. Using JSON schemata, we can define a set of validations that will be checked against the database level every time a document is inserted or updated. This was introduced in MongoDB 3.6. In our case, we are going to use it to make sure that we always have a positive number of items in our inventory.

The `validator` object in the MongoDB shell format is as follows:

```
validator = { validator:
  { $jsonSchema:
  { bsonType: "object",
  required: ["quantity"],
  properties:
  { quantity:
  { bsonType: ["long"],
  minimum: 0,
  description: "we can't have a negative number of items in our
  inventory"
  }
  }
  }
  }
}
```

JSON schema can be used to implement many of the validations that we would usually have in our models in Rails or Django. These keywords are defined in the following table:

Keyword	Validates on Type	Description
enum	All	The enum of allowed values in a field
type	All	The enum of allowed types in a field
minimum/maximum	Numeric	The minimum and maximum values for a numeric field
minLength/maxLength	String	The minimum and maximum lengths allowed for a string field
pattern	String	The regex pattern that the string field must match
required	Objects	The document must contain all the strings defined in the required property array
minItems/maxItems	Arrays	The minimum and maximum lengths of items in the array
uniqueItems	Arrays	If set to true, all items in the array must have unique values
title	N/A	A descriptive title for the developer's use
description	N/A	A description of the developer's use

Table 6.2 – JSON validation types and attributes

Using the JSON schema, we can offload validations from our models to the database layer and/or use MongoDB validations as an additional layer of security on top of web application validations.

To use a JSON schema, we must specify it at the time that we are creating our collection, as follows:

```
> db.createCollection("inventories", validator)
```

Returning to our example, our code will simulate having an inventory of five bull bearings and placing two orders; one by user Alex for two bull bearings, followed by a second order by user Barbara for another four bull bearings.

As expected, the second order will not go through because we don't have enough bull bearings in our inventory to fulfill it. This can be seen in the following code:

```python
from pymongo import MongoClient
from pymongo.errors import ConnectionFailure
from pymongo.errors import OperationFailure


class ECommerce:
def __init__(self):
self.client = MongoClient('localhost', 27017, w='majority')
self.db = self.client.mongo_bank
self.users = self.db['users']
self.carts = self.db['carts']
self.payments = self.db['payments']
self.inventories = self.db['inventories']
# delete any existing data
self.db.drop_collection('carts')
self.db.drop_collection('payments')
self.db.inventories.remove()
# insert new data
self.insert_data()
alex_order_cart_id = self.add_to_cart(1,101,2)
barbara_order_cart_id = self.add_to_cart(2,101,4)
self.place_order(alex_order_cart_id)
self.place_order(barbara_order_cart_id)
def insert_data(self):
self.users.insert_one({'user_id': 1, 'name': 'alex' })
self.users.insert_one({'user_id': 2, 'name': 'barbara'})
```

```python
self.carts.insert_one({'cart_id': 1, 'user_id': 1})
self.db.carts.insert_one({'cart_id': 2, 'user_id': 2})
self.db.payments.insert_one({'cart_id': 1, 'name': 'alex',
'item_id': 101, 'status': 'paid'})
self.db.inventories.insert_one({'item_id': 101, 'description':
'bull bearing', 'price': 100, 'quantity': 5.0})
# Adding an item with quantity to the cart
def add_to_cart(self, user, item, quantity):
# find cart for user
cart_id = self.carts.find_one({'user_id':user})['cart_id']
self.carts.update_one({'cart_id': cart_id}, {'$inc':
{'quantity': quantity}, '$set': { 'item': item} })
return cart_id


# Placing an order
def place_order(self, cart_id):
while True:
try:
with self.client.start_session() as ses:
ses.start_transaction()
cart = self.carts.find_one({'cart_id': cart_id}, session=ses)
item_id = cart['item']
quantity = cart['quantity']
# update payments
self.db.payments.insert_one({'cart_id': cart_id, 'item_id':
item_id, 'status': 'paid'}, session=ses)
# remove item from cart
self.db.carts.update_one({'cart_id': cart_id}, {'$inc':
{'quantity': quantity * (-1)}}, session=ses)
# update inventories
self.db.inventories.update_one({'item_id': item_id}, {'$inc':
{'quantity': quantity*(-1)}}, session=ses)
ses.commit_transaction()
break
except (ConnectionFailure, OperationFailure) as exc:
print("Transaction aborted. Caught exception during
transaction.")
```

```
# If transient error, retry the whole transaction
if exc.has_error_label("TransientTransactionError"):
print("TransientTransactionError, retrying transaction ...")
continue
elif str(exc) == 'Document failed validation':
print("error validating document!")
raise
else:
print("Unknown error during commit ...")
raise
def main():
ECommerce()
if __name__ == '__main__':
main()
```

Let's break down the preceding example into its interesting parts, as follows:

```
def add_to_cart(self, user, item, quantity):
# find cart for user
cart_id = self.carts.find_one({'user_id':user})['cart_id']
self.carts.update_one({'cart_id': cart_id}, {'$inc':
{'quantity': quantity}, '$set': { 'item': item} })
return cart_id
```

The add_to_cart() method doesn't use transactions. This is because we are updating one document at a time, so these are guaranteed to be atomic operations.

Then, in the place_order() method, we start the session, and then subsequently, a transaction within this session. Similar to the previous use case, we need to make sure that we add the session=ses parameter at the end of every operation that we want to be executed in the transaction context:

```
def place_order(self, cart_id):
while True:
try:
with self.client.start_session() as ses:
ses.start_transaction()

...

# update payments
```

```
self.db.payments.insert_one({'cart_id': cart_id, 'item_id':
item_id, 'status': 'paid'}, session=ses)
# remove item from cart
self.db.carts.update_one({'cart_id': cart_id}, {'$inc':
{'quantity': quantity * (-1)}}, session=ses)
# update inventories
self.db.inventories.update_one({'item_id': item_id}, {'$inc':
{'quantity': quantity*(-1)}}, session=ses)
ses.commit_transaction()
break
except (ConnectionFailure, OperationFailure) as exc:
print("Transaction aborted. Caught exception during
transaction.")
# If transient error, retry the whole transaction
if exc.has_error_label("TransientTransactionError"):
print("TransientTransactionError, retrying transaction ...")
continue
elif str(exc) == 'Document failed validation':
print("error validating document!")
raise
else:
print("Unknown error during commit ...")
raise
```

In this method, we are using the retryable transaction pattern. We start by wrapping the transaction context in a `while True` block, essentially making it loop forever. Then, we enclose our transaction in a `try` block that will listen for exceptions.

An exception of the `transient transaction` type, which has the `TransientTransactionError` error label, will result in continued execution in the `while True` block, essentially retrying the transaction from the very beginning. On the other hand, a failed validation or any other error will reraise the exception once it's been logged.

Note

The `session.commitTransaction()` and `session.abortTransaction()` operations will be retried once by MongoDB, regardless of whether we retry the transaction or not.

We don't need to explicitly call `abortTransaction()` in this example since MongoDB will abort it in the face of exceptions.

In the end, our database will look as follows:

```
> db.payments.find()
{ "_id" : ObjectId("5bc307178e72b431c0de385f"), "cart_id" : 1,
"name" : "alex", "item_id" : 101, "status" : "paid" }
{ "_id" : ObjectId("5bc307178e72b431c0de3861"), "cart_id" : 1,
"item_id" : 101, "status" : "paid" }
```

The payment that we just made does not contain the name field, in contrast to the sample payment that we inserted in our database before rolling our transactions:

```
> db.inventories.find()
{ "_id" : ObjectId("5bc303468e72b43118dda074"), "item_id" :
101, "description" : "bull bearing", "price" : 100, "quantity"
: 3 }
```

Our inventory has the correct number of bull bearings, which is three (five minus the two that Alex ordered), as shown in the following code block:

```
> db.carts.find()
{ "_id" : ObjectId("5bc307178e72b431c0de385d"), "cart_id" : 1,
"user_id" : 1, "item" : 101, "quantity" : 0 }
{ "_id" : ObjectId("5bc307178e72b431c0de385e"), "cart_id" : 2,
"user_id" : 2, "item" : 101, "quantity" : 4 }
```

Our carts have the correct quantities. Alex's cart (cart_id=1) has zero items, whereas Barbara's cart (cart_id=2) still has four since we don't have enough bull bearings to fulfill her order. Our payments collection does not have an entry for Barbara's order and the inventory still has three bull bearings in place.

Our database state is consistent and saves lots of time by implementing the abort transaction and reconciliation data logic at the application level.

Continuing with the same example in Ruby, we get the following code block:

```
require 'mongo'

class ECommerce
def initialize
@client = Mongo::Client.new([ '127.0.0.1:27017' ], database:
:mongo_bank)
db = @client.database
@users = db[:users]
```

```ruby
@carts = db[:carts]
@payments = db[:payments]
@inventories = db[:inventories]

# drop any existing data
@users.drop
@carts.drop
@payments.drop
@inventories.delete_many

# insert data
@users.insert_one({ "user_id": 1, "name": "alex" })
@users.insert_one({ "user_id": 2, "name": "barbara" })

@carts.insert_one({ "cart_id": 1, "user_id": 1 })
@carts.insert_one({ "cart_id": 2, "user_id": 2 })

@payments.insert_one({"cart_id": 1, "name": "alex", "item_id":
101, "status": "paid" })
@inventories.insert_one({"item_id": 101, "description": "bull
bearing", "price": 100, "quantity": 5 })

alex_order_cart_id = add_to_cart(1, 101, 2)
barbara_order_cart_id = add_to_cart(2, 101, 4)

place_order(alex_order_cart_id)
place_order(barbara_order_cart_id)
end

def add_to_cart(user, item, quantity)
session = @client.start_session
session.start_transaction
cart_id = @users.find({ "user_id": user}).first['user_id']
@carts.update_one({"cart_id": cart_id}, {'$inc': { 'quantity':
quantity }, '$set': { 'item': item } }, session: session)
session.commit_transaction
cart_id
```

```ruby
end

def place_order(cart_id)
session = @client.start_session
session.start_transaction
cart = @carts.find({'cart_id': cart_id}, session: session).
first
item_id = cart['item']
quantity = cart['quantity']
@payments.insert_one({'cart_id': cart_id, 'item_id': item_id,
'status': 'paid'}, session: session)
@carts.update_one({'cart_id': cart_id}, {'$inc': {'quantity':
quantity * (-1)}}, session: session)
@inventories.update_one({'item_id': item_id}, {'$inc':
{'quantity': quantity*(-1)}}, session: session)
quantity = @inventories.find({'item_id': item_id}, session:
session).first['quantity']
if quantity < 0
session.abort_transaction
else
session.commit_transaction
end
end
end

ECommerce.new
```

Similar to the Python code sample, we are passing the `session: session` parameter along each operation to make sure that we are operating inside the transaction.

Here, we are not using the retry-able transaction pattern. Regardless, MongoDB will retry committing or aborting a transaction once before throwing an exception.

The best practices and limitations of multi-document ACID transactions

At the time of writing, there are some limitations and best practices when developing using MongoDB transactions:

- The transaction timeout is set to 60 seconds.

- As a best practice, any transaction should not try to modify more than 1,000 documents. There is no limitation in reading documents during a transaction.

- The oplog will record a single entry for a transaction, meaning that this is subject to the 16 MB document size limit. This is not such a big problem with transactions that update documents, as only the delta will be recorded in the oplog. It can, however, be an issue when transactions insert new documents, in which case the oplog will record the full contents of the new documents.

- We should add application logic to cater to failing transactions. These could include using retryable writes or executing some business logic-driven action when the error cannot be retried or we have exhausted our retries (usually, this means a custom 500 error).

- DDL operations such as modifying indexes, collections, or databases will get queued up behind active transactions. Transactions trying to access the namespace while a DDL operation is still pending will immediately abort.

- Transactions only work in replica sets. Starting from MongoDB 4.2, transactions will also be available for sharded clusters.

- Use them sparingly; maybe the most important point to consider when developing using MongoDB transactions is that they are not meant as a replacement for good schema design. They should only be used when there is no other way to model our data without them.

Now, let's summarize this chapter.

Summary

In this chapter, we learned about textbook relational database theory on ACID in the context of MongoDB.

Following that, we focused on multi-document ACID transactions and applied them to two use cases using Ruby and Python. We learned about when to use MongoDB transactions and when not to use them, how to use them, their best practices, and their limitations concerning the digital bank and e-commerce use cases.

In the next chapter, we will deal with one of the most commonly used features of MongoDB: aggregation.

7

Aggregation

In *Chapter 6*, *Multi-Document ACID Transactions*, we worked through two use cases of new transactional capabilities using the code for Ruby and Python. In this chapter, we will dive deeper into the aggregation framework, learning how it can be useful. Additionally, we will look at the aggregation operators that are supported by MongoDB.

Following that, we will learn about the time series collections, which were introduced in version 5 and greatly expanded in version 6. Next, we will learn about MongoDB views, which are similar to traditional database materialized views.

Finally, we will discuss the most major aggregation pipeline limitations and bring everything together with a sample aggregation use case.

To learn this information, we will use aggregations to process transaction data from the Ethereum blockchain. The complete source code is available at `https://github.com/PacktPublishing/ Mastering-MongoDB-6.x`.

In this chapter, we will cover the following topics:

- Why aggregation?
- Aggregation options
- Different aggregation operators
- Time series collections
- MongoDB views
- Aggregation limitations
- Optimizing aggregation pipelines
- An aggregation use case

Technical requirements

To follow the examples in this chapter, we will need to install MongoDB, the official PyMongo Python driver v.4.1, and import the data files from `chapter 7` of the book's GitHub repository.

Why aggregation?

The aggregation framework was introduced by MongoDB in version 2.2 (version 2.1 in the development branch). It serves as an alternative to both the MapReduce framework, which is deprecated as of version 5.0, and querying the database directly.

Using the aggregation framework, we can perform `GROUP BY` operations in the server. Therefore, we can project only the fields that are needed in the result set. Using the `$match` and `$project` operators, we can reduce the amount of data passed through the pipeline, resulting in faster data processing.

Self-joins—that is, joining data within the same collection—can also be performed using the aggregation framework, as we will see in our use case.

When comparing the aggregation framework to simply using the queries available via the shell or various other drivers, it is important to remember that there is a use case for both.

For selection and projection queries, it's almost always better to use simple queries, as the complexity of developing, testing, and deploying an aggregation framework operation cannot easily outweigh the simplicity of using built-in commands. Finding documents with (`db.books.find({price: 50} {price: 1, name: 1})`), or without (`db.books.find({price: 50})`) projecting only some of the fields, is simple and fast enough to not warrant the usage of the aggregation framework.

On the other hand, if we want to perform `GROUP BY` and self-join operations using MongoDB, there might be a case for using the aggregation framework. The most important limitation of the `group()` command in the MongoDB shell is that the resulting set has to fit in a document, meaning that it can't be more than 16 MB in size. In addition, the result of any `group()` command can't have more than 20,000 results.

Finally, there is always the option of using the database as data storage and performing complex operations using the application. Sometimes, this can be quick to develop but should be avoided, as it will most likely incur memory, networking, and ultimately performance costs down the road.

In the next section, we will describe the available operators before using them in a real use case.

Aggregation options

Usually, the use case for aggregation is grouping together attribute values from multiple documents or returning a single result after processing a set of documents.

We can also use the aggregation framework to track how data changes over time in a collection.

One example would be calculating the average value for an attribute across all documents of a collection.

We can use either one or more aggregation pipelines, forming a **Directed Acyclic Graph (DAG)** for the calculations or a single-purpose aggregation method for simple one-off calculations.

In earlier versions, we could also use Map Reduce, but as of MongoDB 5.0, the Map-reduce framework has been deprecated and, as such, should not be used for new development.

Existing map reduce code can be rewritten using an aggregation pipeline in multiple stages. The `$accumulator` and `$function` aggregator operators that were introduced in MongoDB 4.4 can be used to define custom aggregation expressions using JavaScript. This should be useful in the somewhat specific case where our mapper and reducer contain custom code that cannot be expressed using an aggregation pipeline. Writing custom code opens up the possibility of code injection vulnerabilities, and this should be weighed as a security risk.

Single-purpose aggregation methods

MongoDB provides three helper aggregation methods that can be used directly from the shell or official drivers to avoid writing code where we want to perform count or distinct operations.

They are `estimatedDocumentCount()`, `count()`, and `distinct()`:

- `count()` is a method wrapper for a combined `$group` and `$sum` aggregation pipeline and, as such, will provide an accurate count of documents at all times.

- `distinct()` will, similarly, perform a distinct count, which can include query or collation options.

- `estimatedDocumentCount()` uses internal MongoDB metadata collection statistics to return a quick estimation of the total number of documents in the collection. It cannot filter based on a query. For example, in the case of an unclean shutdown using the default WiredTiger storage engine, it will return the latest count calculated at the last checkpoint before the shutdown occurred. By default, the checkpoint frequency is 60 seconds. Custom `syncdelay` set at the server level can also affect the delta in seconds between the latest checkpoint and the shutdown.

Finally, using `estimatedDocumentCount()` in sharded clusters might not filter out orphaned documents.

> **Note**
>
> The `estimatedDocumentCount()` method refers to the new mongo shell, mongosh. There might be identical method names in Node.js or other language drivers. Check the documentation, and use the guideline that the older the library is, the greater the odds are that the method is referring to a different definition of estimated count than the previously mentioned calculation that mongosh and compatible drivers use as of MongoDB 5.

Single-purpose aggregation methods can be used as helper methods but should not be, generally, used as a replacement. In the next section, we will learn about the more advanced aggregation operators using pipelines and the recommended usage by MongoDB.

Aggregation operators

In this section, we will learn how to use aggregation operators. Aggregation operators are divided into two categories. Within each stage, we use **expression operators** to compare and process values. Between different stages, we use **aggregation stage operators** to define the data that will get passed on from one stage to the next, as it is considered to be in the same format.

Aggregation pipeline stages

An aggregation pipeline is composed of different stages. These stages are declared in an array and executed sequentially, with the output of every stage being the input of the next one.

The last stage of the pipeline has to be either an $out operator or a $merge operator.

The $out stage has to be the final stage in an aggregation pipeline, outputting data to an output collection that will be completely erased and overwritten if it already exists. The $out operator cannot output to a sharded collection.

Additionally, the $merge stage has to be the final stage in an aggregation pipeline, but it can insert, merge, replace, or keep existing documents. It can also process the documents using a custom update pipeline. It can replace all the documents in a collection but only if the aggregation results output matches all of the existing documents in the collection. The $merge operator can also output to a sharded collection.

At most, the $out and $merge (and $geoNear) stages can only appear once in an aggregation pipeline.

The following list describes the most important aggregation pipeline stages:

- $addFields: This adds new fields to documents and outputs the same number of documents as input with the added fields.
- $bucket: This splits the documents into buckets based on predefined selection criteria and bucket boundaries.
- $bucketAuto: This splits documents into buckets based on predefined selection criteria and attempts to evenly distribute documents among the buckets.
- $collStats: This returns statistics regarding the view or collection.
- $count: This returns a count of the number of documents at this stage of the pipeline.

- $densify: This will create new documents and fill in the gaps in a sequence of documents where the values in the specified field are missing.

- $documents: This will return literal documents from our input values. It can be used for testing or to inject documents into the pipeline programmatically.

- $facet: This combines multiple aggregation pipelines within a single stage.

- $fill: This will populate null and missing field values within the specified documents. We can use the linear regression or **Last Observation Carried Forward** (locf) methods to generate the values.

- $geoNear: This returns an ordered list of documents based on the proximity to a specified field. The output documents include a computed distance field.

- $graphLookup: This recursively searches a collection and adds an array field with the results of the search in each output document.

- $group: This is most commonly used to group by identifier expression and to apply the accumulator expression. It outputs one document per distinct group.

- $indexStats: This returns statistics regarding the indexes of the collection.

- $limit: This limits the number of documents passed on to the next aggregation phase based on predefined criteria.

- $listSessions: This is only available as the first step in the pipeline. It will list active sessions by querying the system.sessions collection. All sessions are initiated in memory local to each node before MongoDB syncs them with the system.sessions collection. We can list in-memory sessions using the $ listLocalSessions operation in the node.

- $lookup: This is used for filtering documents from the input. The input could be documents from another collection in the same database selected by an outer left join or the literal $documents from the input values operator.

- $match: This is used for filtering documents from input based on criteria.

- $merge

- $out: This outputs the documents from this pipeline stage to an output collection by replacing or adding to the documents that already exist in the collection.

- $project: This is used for document transformation, and outputs one document per input document.

- $redact: As a combination of $project and $match, this will redact the selected fields from each document and pass them on to the next stage of the pipeline.

- $replaceRoot: This replaces all existing fields of the input document (including the standard _id field) with the specified fields.

- $replaceWith: This will replace a document with the specified embedded document. The operation replaces all of the existing fields in the input document, including the _id field. Specify a document embedded in the input document to promote the embedded document to the top level. $replaceWith is an alias for the $replaceRoot stage.

- $sample: This randomly selects a specified number of documents from the input.

- $search: This is used to perform a full-text search in the input-specified fields. It can return a snippet of the text including the search term. It can only be the first step in the pipeline; MongoDB Atlas only.

- $searchMeta: MongoDB Atlas only. This is not available in self-hosted clusters.

- $set: This will add new fields to the specified documents. $set will reformat each document passing through the stream and may add new fields to the output documents with both the existing fields and the new ones. It is an alias of the $addFields stage.

- $setWindowFields: This will group input documents into partitions (windows) and apply our operators to all documents in each partition.

- $skip: This skips a certain number of documents, preventing them from passing on to the next stage of the pipeline.

- $sort: This sorts the documents based on criteria.

- $sortByCount: This groups incoming documents based on the value of an expression and computes the count of documents in each bucket.

- $unionWith: This will perform a union of two collections to merge the results from two collections into one result set. It is similar to SQL's UNION ALL operator.

- $unset: This will remove fields from the output documents. $unset is an alias for the $project stage that removes fields.

- $unwind: This transforms an array of *n* elements into *n* documents, mapping each document to one element of the array. The documents are then passed on to the next stage of the pipeline.

Aggregation pipeline expression operators

Within every stage, we can define one or more expression operators to apply our intermediate calculations. This section will focus on these expression operators.

Expression Boolean operators

The Boolean operators are used to pass a true or false value to the next stage of our aggregation pipeline.

We can choose to pass the originating integer, string, or any other type value along, too.

We can use the $and, $or, and $not operators in the same way that we would use them in any programming language.

> **Note**
>
> A Boolean expression will evaluate as false any null, 0, or undefined values.

Expression comparison operators

Comparison operators can be used in conjunction with Boolean operators to construct the expressions that we need to evaluate as true/false for the output of our pipeline's stage.

The most commonly used operators are listed as follows:

- $cmp
- $eq (equal)
- $gt (greater than)
- $gte (greater than or equal)
- $lt
- $lte
- $ne (not equal)

All of these previously mentioned operators return a Boolean value of true or false.

The only operator that doesn't return a Boolean value is $cmp, which returns 0 if the two arguments are equivalent, 1 if the first value is greater than the second, and -1 if the second value is greater than the first.

Set expression and array operators

As with most programming languages, set operations ignore duplicate entries and the order of elements, treating them as sets. The order of results is unspecified, and duplicates will be deduplicated in the result set. Set expressions do not apply recursively to elements of the set, but only to the top level. This means that if a set contains, for example, a nested array, then this array may or may not contain duplicates.

The available set operators are listed as follows:

- $allElementsTrue: This is true if all of the elements in the set evaluate to true.
- $anyElementTrue: This is true if any of the elements in the set evaluate to true.
- $setDifference: This returns the documents that appear in the first input set but not the second.

- $setEquals: This is true if the two sets have the same distinct elements.
- $setIntersection: This returns the intersection of all input sets (that is, the documents that appear in all of the input sets).
- $setIsSubset: This is true if all documents in the first set appear in the second one, even if the two sets are identical.
- $setUnion: This returns the union of all input sets (that is, the documents that appear in at least one of all of the input sets).

The available array operators are listed as follows:

- $arrayElemAt: This is used to retrieve the element at the array index position, starting from zero.
- $arrayToObject: This is used to transform an array into a document.
- $concatArrays: This returns a concatenated array.
- $filter: This returns a subset of the array based on specified criteria.
- $first: This will return the first element of the array. Note that this is different from the $first accumulator operator.
- $firstN: This will return the first N elements of the array. Note that this is different from the $first accumulator operator.
- $in: This returns true if the specified value is in the array; otherwise, it returns false.
- $indexOfArray: This returns the index of the array that fulfills the search criteria. If it does not exist, then it will return -1.
- $isArray: This returns true if the input is an array; otherwise, it returns false.
- $last: This returns the last element of the array. Note that this is different from the $last accumulator operator.
- $lastN: This returns the last N elements of the array. Note that this is different from the $lastN accumulator operator.
- $map: This is similar to JavaScript and the map() function of other languages. This operator will apply the expression to each element of the array and return an array of the resulting values in order. It accepts named parameters.
- $maxN: This returns the largest N values of the array. Note that this is different from the $maxN accumulator operator.
- $minN: This returns the smallest N values of the array. Note that this is different from the $minN accumulator operator.

- `$objectToArray`: This operator will convert a document into an array of documents representing key-value pairs.

- `$range`: This outputs an array containing a sequence of integers according to user-defined inputs.

- `$reduce`: This reduces the elements of an array to a single value according to the specified input.

- `$reverseArray`: This returns an array with the elements in reverse order.

- `$size`: This returns the number of items in the array.

- `$slice`: This returns a subset of the array.

- `$sortArray`: This operator sorts the elements of the array. The array can contain simple values, where we can define 1 for ascending and −1 for descending order. Or it can contain documents, where we can define the field to sort by the order direction in the same way.

- `$zip`: This returns a merged array.

Expression date operators

The date operators are used to extract date information from date fields when we want to calculate statistics based on the day of the week, month, or year using the pipeline:

- `$dateAdd`: This is used to add a number of time units to a date object.

- `$dateDiff`: This is used to get the delta difference between two dates in the defined unit (year, month, and second).

- `$dateFromParts`: This constructs a date object from a set of fields. It can be either year/month-millisecond or an `isoWeekDate` format with year/week/dayOfWeek… /millisecond format.

- `$dateFromString`: This converts a string into a date object according to the defined format. The default format is `%Y-%m-%dT%H:%M:%S.%LZ`.

- `$dateSubtract`: This subtracts a number of time units from a date object. It returns a date.

- `$dateToParts`: This returns a document with the year/month…milliseconds fields from a date object. It can also return an ISO week date format by setting ISO8601 to true.

- `$dateToString`: This will return the string representation of a date.

- `$dateTrunc`: This will truncate a date. We can use `binsize` and `unit` to truncate appropriately. For example, `binsize=2` and `unit=hours` will truncate 11:30 a.m. to 10:00 a.m., truncating to the nearest multiple of 2-hour `bins`.

- `$dayOfMonth`: This is used to return the day of the month within a range of 1 to 31.

- `$dayOfWeek`: This is used to get the day of the week, starting from Sunday(1) to Saturday(7).

- $dayOfYear: This will return the day of the year. It starts from 1, all the way to 366 for a leap year.

- $isoDayOfWeek: This is used to get the day of the week in ISO8601 format, starting from Sunday(1) to Saturday(7).

- $isoWeek: This is used to get the week number in the ISO 8601 date format. This would be an integer from 1 to 53 if the year has 53 weeks. The first week of the year is the week that contains the first Thursday of the year.

- $isoWeekYear: This will return the year number in the ISO 8601 date format according to the date that the last week in the ISO 8601 date format ends with. A year starts on the Monday of the first week of the year and ends on the Sunday of the last week of the year, both inclusive.

- For example, with an ISODate input of Sunday 1/1/2017, this operator will return 2016, as this is the year that ends on the Sunday for this week of the year.

- $second: This will return 0 to 59 or 60 in the case that there is a leap second in the calculation.

- $toDate: This will parse a value to a date object. It will return null on null or missing input. It's a wrapper for `{ $convert: { input: <expression>, to: "date" } }` expression.

- $week: This will return 0 to 53 for the week number. 0 would be the first partial week of the year and 53 the last week of a year with a leap week.

- $year, $month, $hour, $minute, and $milliSecond: These will return the relevant portion of the date in zero-based numbering, except for $month, which returns a value ranging from 1(January) to 12(December).

> **Note**
>
> We can also use the $add and $subtract arithmetic operators with dates.
>
> Here, $add will add a date and a number in milliseconds. Additionally, $subtract with two date arguments will return the difference in milliseconds. When subtracting a number (again, in milliseconds) from a date, the date argument must go first.

Literal expression operator

$literal: We use this operator to pass a value through the pipeline without parsing it. One example of usage would be a string such as $sort that we need to pass along without interpreting it as a field path.

Miscellaneous operators

Miscellaneous operators are various general-purpose operators, as follows:

- $getField: This is useful when we have fields that start with $ or contain . in their name. It will return the value of the field.

- $rand: This will return a random float between 0 and 1. It can contain up to 17 decimal digits and will truncate trailing zeros, so the actual length of the value will vary.

- $sampleRate: Given a float between 0 and 1 inclusively, it will return a number of documents according to the rate. 0 for zero documents returned and 1 to return all documents. The process is non-deterministic and, as such, multiple runs will return a different number of documents. The larger the number of documents in the collection, the bigger chance that sampleRate will converge to the percentage of documents returned. This is a wrapper of the { $match: { $expr: { $lt: [{ $rand: {} }, <sampleRate>] } } } expression.

Object expression operators

The following operators help in working with objects:

- $mergeObjects: This will combine multiple objects within a simple output document.

- $objectToArray: This will convert an object into an array of documents with all key-value pairs.

- $setField: This is useful when we have fields that start with $ or contain . in their name. It will create, update, or delete the specified field in the document.

Expression string operators

Like date operators, string operators are used when we want to transform our data from one stage of the pipeline to the next. Potential use cases include preprocessing text fields to extract relevant information to be used in later stages of a pipeline:

- $concat: This is used to concatenate strings.

- $dateFromString: This is used to convert a DateTime string into a date object.

- $dateToString: This is used to parse a date into a string.

- $indexOfBytes: This is used to return the byte occurrence of the first occurrence of a substring in a string.

- $ltrim: This is used to delete whitespace or the characters from the beginning on the left-hand side of the string.

- $regexFind: This will find the first match of the regular expression in the string. It returns information about the match.

- $regexFindAll: This will find all matches of the regular expression in the string. It returns information about all matches.

- $regexMatch: This will return true or false if it can match the regular expression in the string.

- $replaceOne: This will replace the first instance of a matched string in the input.

- $replaceAll: This will replace all instances of the matched string in the input.

- $rtrim: This will remove whitespace or the specified characters from the end on the right-hand side of a string.

- $strLenBytes: This is the number of bytes in the input string.

- $strcasecmp: This is used in case-insensitive string comparisons. It will return 0 if the strings are equal and 1 if the first string is great; otherwise, it will return -1.

- $substrBytes: This returns the specified bytes of the substring.

- $split: This is used to split strings based on a delimiter. If the delimiter is not found, then the original string is returned.

- $toString: This will convert a value into a string.

- $trim: This will remove whitespace or the specified characters from both the beginning and the end of a string.

- $toLower/$toUpper: These are used to convert a string into all lowercase or all uppercase characters, respectively.

The equivalent methods for code points (a value in Unicode, regardless of the underlying bytes in its representation) are listed as follows:

- $indexOfCP

- $strLenCP

- $substrCP

Text expression operators

We can use a text expression operator to get document-level metadata information such as its textScore, indexKey or searchScore, and searchHighlights snippets when using MongoDB Atlas Search.

- $meta: This will return metadata from the aggregation operation for each document.

Timestamp expression operators

Timestamp expression operators will return values from a timestamp object. Timestamp objects are internally used by MongoDB, and it's recommended that developers use `Date()` instead. Timestamp objects resolve to second precision, with an incrementing ordinal (starting from 1) for multiple objects with the same value:

- `$tsIncrement`: This will return the incrementing ordinal from a timestamp as a Long value.

- `$tsSecond`: This will return the second value from a timestamp as a Long value.

Trigonometry expression operators

Trigonometry expressions help us to perform trigonometric operations on numbers.

The two systems we can convert between are degrees and radians.

Degrees measure an angle in terms of how far we have tilted from our original direction. If we look east and then turn our heads west, we need to tilt our heads by 180 degrees. A full circle has 360 degrees, so if we tilt our head by 360 degrees (or any integer multiple, e.g., 720 or 1,080 degrees), then we will still end up looking east.

A radian measures the angle by the distance traveled. A circle has 2pi (2π) radians, making the 180-degree tilt π radians.

One π is the irrational number 3.14159..., meaning that 1 radian is 180/3.14159 ~= 57.2958... degrees.

Angle values are always input or output in radians.

We can use `$degreesToRadians` and `$radiansToDegrees` to convert between the two systems:

- `$sin`: This will return the sin of a value in radians.

- `$cos`: This will return the cosine of a value in radians.

- `$tan`: This will return the tangent of a value in radians.

- `$asin`: This will return the inverse sin, also known as arc sine, angle of a value in radians.

- `$acos`: This will return the inverse cosine, also known as arc cosine, angle of a value in radians.

- `$atan`: This will return the inverse tangent, also known as arc tangent, angle of a value in radians.

- `$atan2`: With inputs x and y, in that order, it will return the inverse tangent, also known as arc tangent, angle of an x/y expression.

- `$asinh`: This will return the inverse hyperbolic sine, also known as hyperbolic arc sine, of a value in radians.

- $aconh: This will return the inverse hyperbolic cosine, also known as hyperbolic arc cosine, of a value in radians.

- $atanh: This will return the inverse hyperbolic tangent, also known as hyperbolic arc tangent, of a value in radians.

- $sinh: This will return the hyperbolic sine of a value in radians.

- $conh: This will return the hyperbolic cosine of a value in radians.

- $tanh: This will return the hyperbolic tangent of a value in radians.

- $degreesToRadians: This will convert a value into radians from degrees.

- $radiansToDegrees: This will convert a value into degrees from radians.

Type expression operators

Type expression operators in the aggregation framework allow us to inspect and convert between different types:

- $convert: This will convert a value into our target type. For example, { $convert: { input: 1337, to: "bool" } } will output 1 because every nonzero value evaluates to true. Likewise, { $convert: { input: false, to: "int" } } will output 0.

- $isNumber: This will return true if the input expression evaluates to any of the following types: integer, decimal, double, or long. It will return false if the input expression is missing, evaluates to null, or any other BSON type than the truthful ones mentioned earlier.

  $toBool: This will convert a value into its Boolean representation or null if the input value is null.

  It's a wrapper around the { $convert: { input: <expression>, to: "bool" } } expression.

- $toDate: This will convert a value into a date. Numerical values will be parsed as the number of milliseconds since Unix epoch time. Additionally, we can extract the timestamp value from an ObjectId object.

  It's a wrapper around the { $convert: { input: <expression>, to: "date" } } expression.

- $toDecimal: This will convert a value into a Decimal128 value.

  It's a wrapper around the { $convert: { input: <expression>, to: "decimal" } } expression.

- $toDouble: This will convert a value into a Double value.

  It's a wrapper around the { $convert: { input: <expression>, to: "double" } } expression.

- $toInt: This will convert a value into an Integer value.

    It's a wrapper around the { $convert: { input: <expression>, to: "int" } } expression.

- $toLong: This will convert a value into a Long value.

    It's a wrapper around the { $convert: { input: <expression>, to: "long" } } expression.

- $toObject: This will convert a value into an ObjectId value.

- $toString: This will convert a value into a string.

- $type: This will return the type of the input or the "missing" string if the argument is a field that is missing from the input document.

> **Note**
>
> When converting to a numerical value using any of the relevant $to<> expressions mentioned earlier, a string input has to follow the base 10 notation; for example, "1337" is acceptable, whereas its hex representation, "0x539," will result in an error.
>
> The input string can't be a float or decimal.
>
> When converting to a numerical value using any of the relevant $to<> expressions mentioned earlier, a date input will output as a result the number of milliseconds since its epoch time. For example, this could be the Unix epoch time, at the beginning of January 1st, 1970, for the ISODate input.

Expression arithmetic operators

During each stage of the pipeline, we can apply one or more arithmetic operators to perform intermediate calculations. These operators are shown in the following list:

- $abs: This is the absolute value.

- $add: This can add numbers or a number to a date to get a new date.

- $ceil/$floor: These are the ceiling and floor functions, respectively.

- $divide: This is used to divide by two inputs.

- $exp: This raises the natural number, *e*, to the specified exponential power.

- $ln/$log/$log10: These are used to calculate the natural log, the log on a custom base, or a log base ten, respectively.

- $mod: This is the modular value.

- $multiply: This is used to multiply by inputs.

- $pow: This raises a number to the specified exponential power.

- $round: This will round a number to an integer or specified decimal. For example, rounding X.5 will round to the nearest even number, so 11.5 and 12.5 will both round to 12. Also, we can specify a negative round precision number, <place> N, to round the Nth digit to the left-hand side of the decimal point. For example, 1234.56 with N=-2 will round to 1200, as 1234 is closest to 1200 than 1300.

- $sqrt: This is the square root of the input.

- $subtract: This is the result of subtracting the second value from the first. If both arguments are dates, it returns the difference between them. If one argument is a date (this argument has to be the first argument) and the other is a number, it returns the resulting date.

- $trunc: This is used to truncate the result.

Aggregation accumulators

Accumulators are probably the most widely used operators. That's because they allow us to sum, average, get standard deviation statistics, and perform other operations in each member of our group. The following is a list of aggregation accumulators:

- $accumulator: This will evaluate a user-defined accumulator and return its result.

- $addToSet: This will add an element (only if it does not exist) to an array, effectively treating it as a set. It is only available at the group stage.

- $avg: This is the average of the numerical values. It ignores non-numerical values.

- $bottom: This returns the bottom element within a group according to the specified sort order. It is available at the $group and $setWindowFields stages.

- $bottomN: This will return the aggregated bottom N fields within the group as per the defined sort order. We can use it within the $group and $setWindowFields stages.

- $count: This will return the count of documents in a group. Note that this is different from the $count pipeline stage. We can use it within the $group and $setWindowFields stages.

- $first/$last: These are the first and last value that passes through the pipeline stage. They are only available at the group stage.

- $firstN/$lastN: These will return the sum of the first or last N elements within the group. Note that they are different from the $firstN and $lastN array operators.

- $max/$min: These get the maximum and minimum values that pass through the pipeline stage, respectively.

- $maxN: This will return the sum of the first N elements within the group. We can use it within the $group and $setWindowFields stages or as an expression. Note that this is different from the $maxN array operator.

- $push: This will add a new element at the end of an input array. It is only available at the group stage.

- $stdDevPop/$stdDevSamp: These are used to get the population/sample standard deviation in the $project or $match stages.

- $sum: This is the sum of the numerical values. It ignores non-numerical values.

- $top/$topN: This will return the top element or the sum of the top *N* fields within the group respectively, according to the defined sort order.

These accumulators are available in the group or project pipeline phases except where otherwise noted.

Accumulators in other stages

Some operators are available in stages other than the $group stage, sometimes with different semantics. The following accumulators are stateless in other stages and can take one or multiple arguments as input.

These operators are $avg, $max, $min, $stdDevPop, $stdDevSamp, and $sum, and they are covered in the relevant sections of this chapter.

Conditional expressions

Expressions can be used to output different data to the next stage in our pipeline based on Boolean truth tests:

$cond

The $cond phrase will evaluate an expression of the if...then...else format, and depending on the result of the if statement, it will return the value of the then statement or else branches. The input can be either three named parameters or three expressions in an ordered list:

$ifNull

The $ifNull phrase will evaluate an expression and return the first expression if it is not null or the second expression if the first expression is null. Null can be either a missing field or a field with an undefined value:

$switch

Similar to a programming language's switch statement, $switch will execute a specified expression when it evaluates to true and breaks out of the control flow.

Custom aggregation expression operators

We use custom aggregation expression operators to add our own bespoke JavaScript functions to implement custom behavior that's not covered by built-in operators:

- `$accumulator`: This will define a custom accumulator operator. An accumulator maintains its state as documents go through the pipeline stages. For example, we could use an accumulator to calculate the sum, maximum, and minimum values.

- An accumulator can be used in the following stages: `$bucket`, `$bucketAuto`, and `$group`.

- `$function`: This will define a generic custom JavaScript function.

> **Note**
>
> To use these operators, we need to enable server-side scripting. In general, enabling server-side scripting is not recommended by MongoDB due to performance and security concerns.
>
> Implementing custom functions in JavaScript in this context comes with a set of risks around performance and should only be used as a last resort.

Data size operators

We can get the size of the data elements using the following operators.

- `$binarySize`: This will return the size of the input in bytes. The input can be a string or binary data. The binary size per character can vary. For example, English alphabet characters will be encoded using 1 byte, Greek characters will use 2 bytes per character, and ideograms such as might use 3 or more bytes.

- `$bsonSize`: This will take a document's representation as input, for example, an object, and return the size, in bytes, of its BSON-encoded representation.

- The null input will return a null output and any other input than an object will result in an error.

Variable expression operators

We can use a variable expression operator to define a variable within a subexpression in the aggregation stage using the `$let` operator. The `$let` operator defines variables for use within the scope of a subexpression and returns the result of the subexpression. It accepts named parameters and any number of argument expressions.

Window operators

We can define a span of documents from a collection by using an expression in the `partitionBy` field during the `$setWindowFields` aggregation stage.

Partitioning our collection's documents in spans will create a "window" into our data, resembling window functions in relational databases:

- $addToSet: This will apply an input expression to each document of the partition and return an array of all of the unique values.

- $avg: This will return the average of the input expression values, for each document of the partition.

- $bottom: This will return the bottom element of the group according to the specified input order.

- $bottomN: This will return the sum of the bottom *N* fields within the group according to the specified input order.

- $count: This will return the number of documents in the group or window.

- $coVariancePop: This will return the covariance population of two numerical expressions.

- $covarianceSamp: This will return the sample covariance of two numerical expressions.

- $denseRank: This will return the position of the document (rank) in the $setWindowFields partition. Tied documents will receive the same rank. The rank numbering is consecutive.

- $derivative: This will return the first derivative, that is, the average rate of change within the input window.

- $documentNumber: This will return the position of the document in the partition. In contrast with the $denseRank operator, ties will result in consecutive numbering.

- $expMovingAvg: With a numerical expression as input, this operator will return the exponential moving average calculation.

- $first: This will apply the input expression to the first document in the group or window and return the resulting value.

- $integral: Similar to the mathematical integral, this operator will return the calculation of an area under a curve.

- $last: This will apply the input expression to the last document in the group or window and return the resulting value.

- $linearFill: This will use linear interpolation based on surrounding field values to fill null and missing fields in a window.

- $locf: This will use the last non-null value for a field to set values for subsequent null and missing fields in a window.

- $max: This will apply the input expression to each document and return the maximum value.

- $min: This will apply the input expression to each document and return the minimum value.

- $minN: This will return the sum of the *N* minimum valued elements in the group. This is different from the $minN array operator.

- $push: This will apply the input expression to each document and return a result array of values.

- $rank: This will return the document position, meaning the rank of this document relative to the rest of the documents in the $setWindowFields stage partition.

- $shift: This will return the value from an expression applied to a document in a specified position relative to the current document in the $setWindowFields stage partition.

- $stdDevPop: This will apply the input numerical expression to each document in the window and return the population's standard deviation.

- $stdDevSamp: This will apply the input numerical expression to each document in the window and return the population's sample standard deviation.

- $sum: This will apply the input numerical expression to each document in the window and return the sum of their values.

- $top: This will return the top element within the group, respecting the specified sorting order.

- $topN: This will return the sum of the top *N* element within the group, respecting the specified sorting order.

Type conversion operators

Introduced in MongoDB 4.0, type conversion operators allow us to convert a value into a specified type. The generic syntax of the command is as follows:

```
{
{                                                              input:
<expression>,
        to: <type expression>,
        onError: <expression>,   // Optional.
        onNull: <expression>     // Optional.
    }
}
```

In the preceding syntax, input and to (the only mandatory arguments) can be any valid expression. For example, in its simplest form, we could have the following:

```
$convert: { input: "true", to: "bool" }
```

This converts a string with the value of true into the Boolean value of true.

Again, the `onError` phrase can be any valid expression and specifies the value that MongoDB will return if it encounters an error during conversion, including unsupported type conversions. Its default behavior is to throw an error and stop processing.

Additionally, the `onNull` phrase can be any valid expression and specifies the value that MongoDB will return if the input is null or missing. The default behavior is to return null.

MongoDB also provides some helper functions for the most common `$convert` operations. These functions are listed as follows:

- `$toBool`
- `$toDate`
- `$toDecimal`
- `$toDouble`
- `$toInt`
- `$toLong`
- `$toObjectId`
- `$toString`

These are even simpler to use. We could rewrite the previous example as the following:

```
{ $toBool: "true" }
```

After going through the extensive list of aggregation operators, in the next section, we will learn about a new feature of MongoDB 5: time series collections.

Time series collections

A time series collection is a special type of collection that is used to collect data measurements over a period of time.

For example, time series collection use cases can include storing **Internet of Things (IoT)** sensor readings, weather readings, and stock price data.

A time series collection needs to be created as such, and we cannot change a collection type into a time series one. Migrating data from a generic purpose collection to a time series one can be done using a custom script or MongoDB's own Kafka connector for performance and stability.

To create a time series collection, we need to specify the following fields. In this context, a data point might refer to a sensor reading or the stock price at a specific point in time:

- `timeField`: This field is mandatory and is the field that stores the timestamp of the data point. It must be a `Date()` object.

- `metaField`: This field is optional and is used to store metadata for the data point. The metadata field can be an embedded document and should be used to add any data that is uniquely identifying our data point.

- `granularity`: This field is optional and is used to help MongoDB optimize the storage of our data points. It can be set to "seconds," "minutes," or "hours" and should be set to the nearest match between our data points' consecutive readings.

- `expireAfterSeconds`: This field is optional. We can set it to the number of seconds that we would like MongoDB to automatically delete the data points after they are created.

Time series collections create and use an internal index that is not listed in the `listIndexes` command. However, we can manually create additional (namely, secondary) indexes if needed. MongoDB 6.0 has expanded the number of available secondary indexes, including compound indexes, partial indexes with `$in`, `$or`, or `$geoWithin`, and 2dsphere indexes.

Usually, time series collections use cases refer to data that we insert once, read many times, but rarely get updated. We would rarely need to update a weather sensor reading from last week unless we wanted to check with other sources and update our readings with the correct values.

Due to their unique nature, update and delete operations in a time series type of collection are constrained. Updates and deletes need to operate with `multi:true` and `justOne:false` to affect multiple documents, and the query to update or delete might only match the `metaField` subdocument fields. We can only modify the `metaField` subdocument using the update command, and the update cannot be an `upsert` operation as this would create a new document instead of updating an existing one. Transactions cannot write into time series collections.

Time series collections use the `zstd` compression algorithm instead of the default snappy "for generic purpose" collections. This is configurable at collection creation time using the `block_compressor=snappy|zlib|zstd` parameter.

We can define the `storageEngine`, `indexOptionDefaults`, `collation`, and `writeConcern` parameters for a time series collection in the same way as a general-purpose collection.

> **Note**
>
> A time series' maximum document size is limited to 4 MB instead of the global 16 MB.
>
> Time series are optimized for write once, read many with sporadic updates/deletes, and this reflects the limitations in updates/deletes in favor of improved read/write performance.
>
> We can shard a time series collection using `timeField` or `metaField`. In this scenario, `timeField` is a timestamp, so using it on its own will result in all writes going to the last shard. Therefore, it's a good practice to combine `timeField` with one or more fields from `metaField`.

In certain use cases, the time series collections can be really useful. MongoDB 6.0 is expanding on secondary and compound indexes to make them even more useful. It is also adding performance improvements in sorted queries and when fetching the most recent entry from a time series collection. In the next section, we will go through another recent addition to MongoDB from the relational database world: the views functionality.

MongoDB views

Views and materialized views are essential parts of database applications, and they are supported in nearly all relational database management systems.

The main difference between the two is that a view will return the results by querying the underlying data at the time of the query, whereas a materialized view stores the view data in a distinct dataset, meaning that the result set might be outdated by the time of the query.

We can create a MongoDB view using the shell or our programming language's driver with the following parameters:

- `viewName`: This field is mandatory. It refers to the name of the view.
- `viewOn`: This field is mandatory. It refers to the name of the collection that will be used as the data source for the view data.
- `pipeline`: This field is mandatory. The pipeline will execute every time we query the view. The pipeline cannot include the `$out` or `$merge` stages during any stage, including any nested pipelines.
- `collation`: This field is optional. We can define custom language-specific rules for string comparison; for uppercase, lowercase, or accent marks.

> **Note**
>
> Querying a view will execute the aggregation pipeline every time. Its performance is limited by the performance of the aggregation pipeline stages and all pipeline limitations will be applied.

We can create an on-demand materialized view using the aggregation framework and, more specifically, the $merge stage. The $merge operator allows us to keep existing collection documents and merge our output result set into it. We can execute the pipeline periodically to refresh the collection's data that will be used for querying our materialized view.

In the next section, we will discuss the most important aggregation pipeline limitations that we need to keep in mind when designing pipelines.

Limitations

The aggregation pipeline can output results in the following three distinct ways:

- Inline as a document containing the result set
- In a collection
- Returning a cursor to the result set

Inline results are subject to a BSON maximum document size of 16 MB, meaning that we should use this only if our final result is of a fixed size. An example of this would be outputting ObjectId values of the top five most ordered items from an e-commerce site.

A contrary example to this would be outputting the top 1,000 most ordered items, along with the product information, including a description and various other fields of variable size.

Outputting results into a collection is the preferred solution if we want to perform further processing of the data. We can either output into a new collection or replace the contents of an existing collection. The aggregation output results will only be visible once the aggregation command succeeds; otherwise, it will not be visible at all.

> **Note**
>
> The output collection cannot be a sharded or capped collection (as of version 3.4). If the aggregation output violates the indexes (including the built-in index of the unique ObjectId value per document) or document validation rules, the aggregation will fail.

Each pipeline stage can have documents exceeding the 16 MB limit as these are handled by MongoDB internally. However, each pipeline stage can use more than 100 MB of memory by default starting from MongoDB 6.0, flushing excess data to disk at the expense of performance. If we want to return an error instead, we must set {allowDiskUseByDefault: false} at the mongod server level. We can also override this behavior in individual operations using the {allowDiskUse: true|false} operator. The $graphLookup operator does not support datasets that are over 100 MB and will ignore any setting in allowDiskUse.

These limitations should be taken into account when designing pipelines and must be tested thoroughly before being put into production workloads. In the next section, we will discuss how we can optimize aggregation pipelines.

Optimizing aggregation pipelines

When we submit the aggregation pipeline for execution, MongoDB might reorder and group execution stages to improve performance.

We can inspect these optimizations by including the `explain` option in the aggregation execution command in the shell or our programming language's driver.

Further to the optimizations that MongoDB will perform independently, we should aim to design our aggregation commands with an eye on limiting the number of documents that need to pass from one stage to the next, as early as possible.

This can be done in two ways.

The first way is by using indexes to improve the querying speed in every step of the aggregation pipeline. The rules are as follows:

- The planner will use an index in a `$match` stage if `$match` is the first stage in the pipeline. The planner will use an index in the `$sort` stage if `$sort` is not preceded by a `$group`, `$project`, or `$unwind` stage.

- The planner might use an index in the `$group` stage, but just to find the first document in each group if all the following conditions are satisfied:

  - `$group` follows a `$sort` stage, which sorts the grouped fields by key.

  - Previously, we added an index to the grouped field matching the sorting order.

  - `$first` is the only accumulator in `$group`.

- The planner will use a geospatial index in the `$geoNear` stage if `$geoNear` is the first stage in the pipeline.

The second way to do this independently is by designing our pipeline stages to filter out as many documents as early as possible in the aggregation pipeline stages.

Aggregation use case

In this rather lengthy section, we will use the aggregation framework to process data from the Ethereum blockchain.

Using our Python code, we have extracted data from Ethereum and loaded it into our MongoDB database. The relationship of the blockchain to our database is shown in the following diagram:

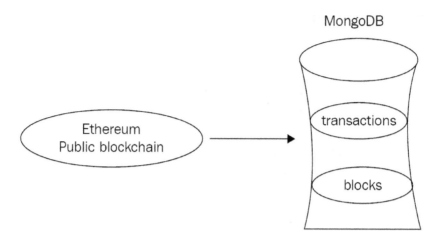

Figure 7.1: A MongoDB database interacting with the Ethereum public blockchain

Our data resides in two collections: **blocks** and **transactions**.

A sample block document has the following fields:

- The number of transactions
- The number of contracted internal transactions
- The block hash
- The parent block hash
- The mining difficulty
- The amount of gas used
- The block height

The following code shows the output data from a block:

```
> db.blocks.findOne()
{
"_id" : ObjectId("595368fbcedea89d3f4fb0ca"),
"number_transactions" : 28,
"timestamp" : NumberLong("1498324744877"),
"gas_used" : 4694483,
"number_internal_transactions" : 4,
"block_hash" : "0x89d235c4e2e4e4978440f3cc1966f1ffb343b9b5c-
fec9e5cebc331fb810bded3",
```

```
"difficulty" : NumberLong("882071747513072"),
"block_height" : 3923788
}
```

A sample transaction document has the following fields:

- The transaction hash
- The block height that it belongs to
- The from hash address
- The to hash address
- The transaction value
- The transaction fee

The following code shows the output data from a transaction:

```
> db.transactions.findOne()
{
"_id" : ObjectId("59535748cedea89997e8385a"),
"from" : "0x3c540be890df69eca5f0099bbedd5d667bd693f3",
"txfee" : 28594,
"timestamp" : ISODate("2017-06-06T11:23:10Z"),
"value" : 0,
"to" : "0x4b9e0d224dabcc96191cace2d367a8d8b75c9c81",
"txhash" : "0xf205991d937bcb60955733e760356070319d95131a-
2d9643e3c48f2dfca39e77",
"block" : 3923794
}
```

Sample data for our database is available on GitHub at https://github.com/
PacktPublishing/Mastering-MongoDB-6.x.

As curious developers who are using this novel blockchain technology, we want to analyze the Ethereum transactions. We are especially keen to do the following:

- Find the top 10 addresses that the transactions originate from.
- Find the top 10 addresses that the transactions end in.
- Find the average value per transaction, with statistics concerning the deviation.

- Find the average fee required per transaction, with statistics concerning the deviation.
- Find the time of day that the network is more active according to the number or value of transactions.
- Find the day of the week in which the network is more active according to the number or value of transactions.

We will find the top 10 addresses that the transactions originate from. To calculate this metric, first, we will count the number of occurrences with a 1 count for each one, group them by the value of the from field, and output them into a new field called count.

After this, we will sort by the value of the count field in descending (-1) order, and finally, we will limit the output to the first 10 documents that pass through the pipeline. These documents are the top 10 addresses that we are looking for.

The following is some sample Python code:

```python
def top_ten_addresses_from(self):
pipeline = [
{"$group": {"_id": "$from", "count": {"$sum": 1}}},
{"$sort": SON([("count", -1)])},
{"$limit": 10},
]
result = self.collection.aggregate(pipeline)
for res in result:
print(res)
```

The output of the preceding code is as follows:

```
{u'count': 38, u'_id': u'miningpoolhub_1'}
{u'count': 31, u'_id': u'Ethermine'}
{u'count': 30, u'_id':
u'0x3c540be890df69eca5f0099bbedd5d667bd693f3'}
{u'count': 27, u'_id':
u'0xb42b20ddbeabdc2a288be7ff847ff94fb48d2579'}
{u'count': 25, u'_id': u'ethfans.org'}
{u'count': 16, u'_id': u'Bittrex'}
{u'count': 8, u'_id':
u'0x009735c1f7d06faaf9db5223c795e2d35080e826'}
{u'count': 8, u'_id': u'Oraclize'}
{u'count': 7, u'_id':
```

```
u'0x1151314c646ce4e0efd76d1af4760ae66a9fe30f'}
{u'count': 7, u'_id':
u'0x4d3ef0e8b49999de8fa4d531f07186cc3abe3d6e'}
```

Now, we will find the top 10 addresses where the transactions end. As we did with `from`, the calculation for the `to` addresses is exactly the same, only grouping using the `to` field instead of `from`, as shown in the following code:

```
def top_ten_addresses_to(self):
pipeline = [
{"$group": {"_id": "$to", "count": {"$sum": 1}}},
{"$sort": SON([("count", -1)])},
{"$limit": 10},
]
result = self.collection.aggregate(pipeline)
for res in result:
print(res)
```

The output of the preceding code is as follows:

```
{u'count': 33, u'_id':
u'0x6090a6e47849629b7245dfa1ca21d94cd15878ef'}
{u'count': 30, u'_id':
u'0x4b9e0d224dabcc96191cace2d367a8d8b75c9c81'}
{u'count': 25, u'_id':
u'0x69ea6b31ef305d6b99bb2d4c9d99456fa108b02a'}
{u'count': 23, u'_id':
u'0xe94b04a0fed112f3664e45adb2b8915693dd5ff3'}
{u'count': 22, u'_id':
u'0x8d12a197cb00d4747a1fe03395095ce2a5cc6819'}
{u'count': 18, u'_id':
u'0x91337a300e0361bddb2e377dd4e88ccb7796663d'}
{u'count': 13, u'_id':
u'0x1c3f580daeaac2f540c998c8ae3e4b18440f7c45'}
{u'count': 12, u'_id':
u'0xeef274b28bd40b717f5fea9b806d1203daad0807'}
{u'count': 9, u'_id':
u'0x96fc4553a00c117c5b0bed950dd625d1c16dc894'}
{u'count': 9, u'_id':
u'0xd43d09ec1bc5e57c8f3d0c64020d403b04c7f783'}
```

Let's find the average value per transaction, with statistics concerning the standard deviation. In this example, we are using the $avg and $stdDevPop operators of the values of the value field to calculate the statistics for this field. Using a simple $group operation, we output a single document with the ID of our choice (here, it is value) and averageValues, as shown in the following code:

```
def average_value_per_transaction(self):
pipeline = [
{"$group": {"_id": "value", "averageValues": {"$avg":
"$value"}, "stdDevValues": {"$stdDevPop": "$value"}}},
]
result = self.collection.aggregate(pipeline)
for res in result:
print(res)
```

The output of the preceding code is as follows:

```
{u'averageValues': 5.227238976440972, u'_id': u'value',
u'stdDevValues': 38.90322689649576}
```

Let's find the average fee required per transaction, returning statistics that showcase the deviation. Average fees are similar to average values, replacing $value with $txfee, as shown in the following code:

```
def average_fee_per_transaction(self):
pipeline = [
{"$group": {"_id": "value", "averageFees": {"$avg": "$txfee"},
"stdDevValues": {"$stdDevPop": "$txfee"}}},
]
result = self.collection.aggregate(pipeline)
for res in result:
print(res)
```

The output of the preceding code snippet is as follows:

```
{u'_id': u'value', u'averageFees': 320842.0729166667,
u'stdDevValues': 1798081.7305142984}
```

We find the time of day that the network is more active according to the number or value of transactions at specific times.

To find out the most active hour for transactions, we use the $hour operator to extract the hour field from the ISODate() field in which we stored our datetime values and called timestamp, as shown in the following code:

```
def active_hour_of_day_transactions(self):
pipeline = [
{"$group": {"_id": {"$hour": "$timestamp"}, "transactions":
{"$sum": 1}}},
{"$sort": SON([("transactions", -1)])},
{"$limit": 1},
]
result = self.collection.aggregate(pipeline)
for res in result:
print(res)
```

The output is as follows:

```
{u'_id': 11, u'transactions': 34}
```

The following code will calculate the sum of transaction values for the most active hour of the day:

```
def active_hour_of_day_values(self):
pipeline = [
{"$group": {"_id": {"$hour": "$timestamp"}, "transaction_
values": {"$sum": "$value"}}},
{"$sort": SON([("transactions", -1)])},
{"$limit": 1},
]
result = self.collection.aggregate(pipeline)
for res in result:
print(res)
```

The output of the preceding code is as follows:

```
{u'transaction_values': 33.17773841, u'_id': 20}
```

Let's find the day of the week that the network is more active according to the number of transactions or the value of transactions. As we did with the hour of the day, we use the $dayOfWeek operator to extract the day of the week from the ISODate() objects, as shown in the following code. Days are numbered ranging from one for Sunday to seven for Saturday, following the US convention:

```
def active_day_of_week_transactions(self):
pipeline = [
{"$group": {"_id": {"$dayOfWeek": "$timestamp"},
"transactions": {"$sum": 1}}},
{"$sort": SON([("transactions", -1)])},
{"$limit": 1},
]
result = self.collection.aggregate(pipeline)
for res in result:
print(res)
```

The output of the preceding code is as follows:

```
{u'_id': 3, u'transactions': 92}
```

The following code will calculate the sum of transaction values for the most active day of the week:

```
def active_day_of_week_values(self):
    pipeline = [
        {"$group": {"_id": {"$dayOfWeek": "$timestamp"},
"transaction_values": {"$sum": "$value"}}},
        {"$sort": SON([("transactions", -1)])},
        {"$limit": 1},
    ]
    result = self.collection.aggregate(pipeline)
for res in result:
print(res)
The output of the preceding code is as follows:
{u'transaction_values': 547.62439312, u'_id': 2}
```

The aggregations that we calculated can be described in the following diagram:

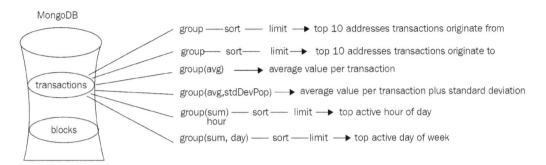

Figure 7.2: Aggregations performed in the transactions collection

In terms of blocks, we would like to know the following:

- The average number of transactions per block, for both the total number of overall transactions and the total number of contracted internal transactions.

- The average amount of gas used per block.

- The average amount of gas used per transaction to a block. Is there a window of opportunity in which to submit my smart contract in a block?

- The average level of difficulty per block and how it deviates.

- The average number of transactions per block, both in total and also in contracted internal transactions.

Averaging over the `number_transactions` field, we can get the number of transactions per block, as shown in the following code:

```
def average_number_transactions_total_block(self):
    pipeline = [
        {"$group": {"_id": "average_transactions_per_block",
"count": {"$avg": "$number_transactions"}}},
    ]
    result = self.collection.aggregate(pipeline)
    for res in result:
        print(res)
```

The output of the preceding code is as follows:

```
{u'count': 39.458333333333336, u'_id': u'average_transactions_
per_block'}
```

Whereas with the following code, we can get the average number of internal transactions per block:

```
def average_number_transactions_internal_block(self):
    pipeline = [
        {"$group": {"_id": "average_transactions_
internal_per_block", "count": {"$avg": "$number_internal_
transactions"}}},
    ]
    result = self.collection.aggregate(pipeline)
    for res in result:
        print(res)
```

The output of the preceding code is as follows:

```
{u'count': 8.0, u'_id': u'average_transactions_internal_per_
block'}
```

The average amount of gas used per block can be acquired as follows:

```
def average_gas_block(self):
    pipeline = [
        {"$group": {"_id": "average_gas_used_per_block",
                    "count": {"$avg": "$gas_used"}}},
    ]
    result = self.collection.aggregate(pipeline)
    for res in result:
        print(res)
```

The output is as follows:

```
{u'count': 2563647.9166666665, u'_id': u'average_gas_used_per_
block'}
```

The average difficulty per block and how it deviates can be acquired as follows:

```
def average_difficulty_block(self):
    pipeline = [
        {"$group": {"_id": "average_difficulty_per_block",
                    "count": {"$avg": "$difficulty"},
"stddev": {"$stdDevPop": "$difficulty"}}},
    ]
```

```
            result = self.collection.aggregate(pipeline)
            for res in result:
                print(res)
```

The output is as follows:

```
{u'count': 881676386932100.0, u'_id': u'average_difficulty_per_
block', u'stddev': 446694674991.6385}
```

Our aggregations are described in the following diagram:

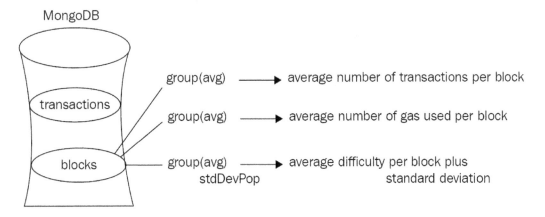

Figure 7.3: Aggregations performed in the blocks collection

Now that we have the basic statistics calculated, we want to up our game and find out more information about our transactions. Through our sophisticated machine learning algorithms, we have identified some of the transactions as either a scam or an **initial coin offering** (**ICO**), or maybe both.

In these documents, we have marked the attributes in an array called tags, as follows:

```
{
  "_id" : ObjectId("59554977cedea8f696a416dd"),
  "to" : "0x4b9e0d224dabcc96191cace2d367a8d8b75c9c81",
  "txhash" : "0xf205991d937bcb60955733e760356070319d95131a-
2d9643e3c48f2dfca39e77",
  "from" : "0x3c540be890df69eca5f0099bbedd5d667bd693f3",
  "block" : 3923794,
  "txfee" : 28594,
  "timestamp" : ISODate("2017-06-10T09:59:35Z"),
  "tags" : [
```

```
"scam",
"ico"
],
"value" : 0
}
```

Now we want to get the transactions from June 2017, remove the _id field, and produce different documents according to the tags that we have identified. So, in our example, we would output two documents in our new collection, scam_ico_documents, for separate processing.

The way to do this is via the aggregation framework, as shown in the following code:

```
def scam_or_ico_aggregation(self):
  pipeline = [
  {"$match": {"timestamp": {"$gte": datetime.datetime(2017,6,1),
"$lte": datetime.datetime(2017,7,1)}}},
  {"$project": {
  "to": 1,
  "txhash": 1,
  "from": 1,
  "block": 1,
  "txfee": 1,
  "tags": 1,
  "value": 1,
  "report_period": "June 2017",
  "_id": 0,
  }

  },
  {"$unwind": "$tags"},
  {"$out": "scam_ico_documents"}
  ]
  result = self.collection.aggregate(pipeline)
  for res in result:
  print(res)
```

Here, we have the following four distinct steps in our aggregation framework pipeline:

1. Using $match, we only extract documents that have a timestamp field value of June 01, 2017.

2. Using $project, we add a new report_period field with a value of June 2017 and remove the _id field by setting its value to 0. We keep the rest of the fields intact by using the value of 1, as shown in the preceding code.

3. Using $unwind, we output one new document per tag in our $tags array.

4. Finally, using $out, we output all of our documents to a new scam_ico_documents collection.

Since we used the $out operator, we will get no results in the command line. If we comment out {"$out": "scam_ico_documents"}, we get documents that look like the following:

```
{u'from': u'miningpoolhub_1', u'tags': u'scam', u'report_pe-
riod': u'June 2017', u'value': 0.52415349, u'to': u'0xdaf112b-
cbd38d231b1be4ae92a72a41aa2bb231d', u'txhash': u'0xe11ea11d-
f4190bf06cbdaf19ae88a707766b007b3d9f35270cde37ceccba9a5c',
u'txfee': 21.0, u'block': 3923785}
```

The final result in our database will look like this:

```
{
  "_id" : ObjectId("5955533be9ec57bdb074074e"),
  "to" : "0x4b9e0d224dabcc96191cace2d367a8d8b75c9c81",
  "txhash" : "0xf205991d937bcb60955733e760356070319d95131a-
2d9643e3c48f2dfca39e77",
  "from" : "0x3c540be890df69eca5f0099bbedd5d667bd693f3",
  "block" : 3923794,
  "txfee" : 28594,
  "tags" : "scam",
  "value" : 0,
  "report_period" : "June 2017"
}
```

Now that we have documents that are clearly separated in the scam_ico_documents collection, we can perform further analysis pretty easily. An example of this analysis would be to append more information on some of these scammers. Luckily, our data scientists have come up with some additional information that we have extracted into a new scam_details collection, which looks like this:

```
{
  "_id" : ObjectId("5955510e14ae9238fe76d7f0"),
  "scam_address" : "0x3c540be890df69eca5f0099bbedd5d667bd693f3",
  Email_address": example@scammer.com"
}
```

Now, we can create a new aggregation pipeline job to join scam_ico_documents with the scam_details collection and output these extended results into a new collection, called scam_ico_documents_extended, as follows:

```
def scam_add_information(self):
  client = MongoClient()
  db = client.mongo_book
  scam_collection = db.scam_ico_documents
  pipeline = [
  {"$lookup": {"from": "scam_details", "localField": "from",
"foreignField": "scam_address", "as": "scam_details"}},
  {"$match": {"scam_details": { "$ne": [] }}},
  {"$out": "scam_ico_documents_extended"}
  ]
  result = scam_collection.aggregate(pipeline)
  for res in result:
  print(res)
```

Here, we are using the following three-step aggregation pipeline:

5. Use the $lookup command to join data from the scam_details collection and the scam_address field with data from our local collection (scam_ico_documents) based on the value from the local collection attribute, from, being equal to the value in the scam_details collection's scam_address field. If they are equal, then the pipeline adds a new field to the document called scam_details.

6. Next, we only match the documents that have a scam_details field—the ones that matched with the lookup aggregation framework step.

7. Finally, we output these documents into a new collection called scam_ico_documents_extended.

These documents will now look like the following:

```
> db.scam_ico_documents_extended.findOne()
 {
 "_id" : ObjectId("5955533be9ec57bdb074074e"),
 "to" : "0x4b9e0d224dabcc96191cace2d367a8d8b75c9c81",
 "txhash" : "0xf205991d937bcb60955733e760356070319d95131a-
2d9643e3c48f2dfca39e77",
 "from" : "0x3c540be890df69eca5f0099bbedd5d667bd693f3",
 "block" : 3923794,
```

```
"txfee" : 28594,
"tags" : "scam",
"value" : 0,
"report_period" : "June 2017",
"scam_details_data" : [
{
"_id" : ObjectId("5955510e14ae9238fe76d7f0"),
"scam_address" : "0x3c540be890df69eca5f0099bbedd5d667bd693f3",
email_address": example@scammer.com"
}]}
```

Using the aggregation framework, we have identified our data and can process it rapidly and efficiently.

The previous steps can be summed up in the following diagram:

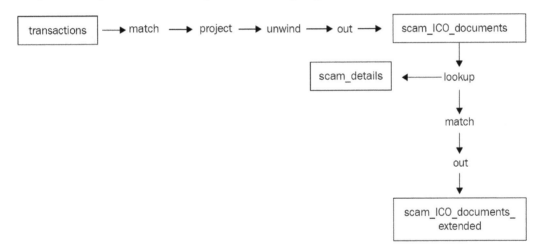

Figure 7.4: The overall data flow of our use case

Summary

In this chapter, we dived deep into the aggregation framework. We discussed why and when we should use aggregation as opposed to simply using MapReduce or querying the database. We went through a vast array of options and functionalities for aggregation.

We discussed the aggregation stages and the various operators, such as Boolean operators, comparison operators, set operators, array operators, date operators, string operators, expression arithmetic operators, aggregation accumulators, conditional expressions, and variables, along with the literal parsing data type operators.

Using the Ethereum use case, we went through aggregation with working code and learned how to approach an engineering problem to solve it.

Finally, we learned about the limitations that the aggregation framework currently has and when to avoid them.

In the next chapter, we will move on to the topic of indexing and learn how to design and implement performant indexes for our read and write workloads.

8
Indexing

This chapter will explore one of the most important properties of any database: indexing. Similar to book indexes, database indexes allow for quicker data retrieval. In the **relational database management system (RDBMS)** world, indexes are widely used (and sometimes abused) to speed up data access. In MongoDB, indexes play an integral part in schema design and query design. MongoDB supports a wide array of indexes, which you will learn about in this chapter, including single-field, compound, multi-key, geospatial, hashed, and partial. In addition to reviewing the different types of indexes, we will show you how to build and manage indexes for single-server deployments, as well as complex sharded environments.

In this chapter, we will cover the following topics:

- Index internals
- Types of indexes
- Building and managing indexes
- Efficient usage of indexes

Index internals

In most cases, indexes are variations of the **B-tree** data structure. Invented by Rudolf Bayer and Ed McCreight in 1971 while they were working at Boeing research labs, the B-tree data structure allows for searches, sequential access, inserts, and deletes to be performed in logarithmic time. The **logarithmic time** property stands for both the average case performance and the worst possible performance, and it is a great property when applications cannot tolerate unexpected variations in performance behavior.

To further illustrate how important the logarithmic time part is, we will show you the Big O complexity chart, which is from `http://bigocheatsheet.com/`:

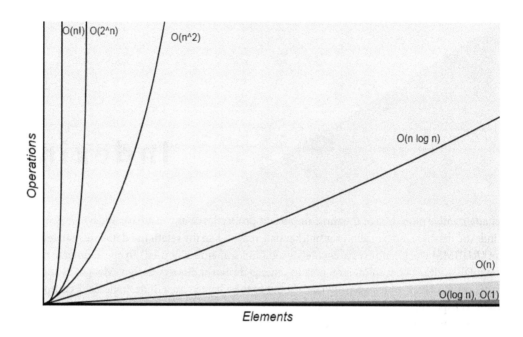

Figure 8.1 – Algorithmic complexity visualized

In this diagram, you can see logarithmic time performance as a flat line, parallel to the *x* axis of the diagram. As the number of elements increases, constant time (**O(n)**) algorithms perform worse, whereas quadratic time algorithms (**O(n^2)**) go off the chart. For an algorithm that we rely on to get our data back to us as quickly as possible, time performance is of the utmost importance.

Another interesting property of a B-tree is that it is self-balancing, meaning that it will self-adjust to always maintain these properties. Its precursor and closest relative is the binary search tree, a data structure that only allows two children for each parent node.

Schematically, a B-tree looks like the following diagram, which can also be seen at `https://commons.wikimedia.org/w/index.php?curid=11701365`:

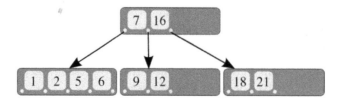

Figure 8.2 – An internal representation of a B-tree

In the preceding diagram, we have a parent node with values **7** and **16** pointing to three child nodes.

If we search for value **9**, knowing that it's greater than **7** and smaller than **16**, we'll be directed to the middle child node, which contains the value, straight away.

Thanks to this structure, we are approximately halving our search space with every step, ending in a *log n* time complexity. Compared to sequentially scanning through every element, halving the number of elements with each and every step increases our gains exponentially as the number of elements we have to search through increases.

Index types

MongoDB offers a vast array of index types for different needs. In the following sections, we will identify the different types and the needs that each one of them fulfills.

Single-field indexes

The most common and simple type of index is the single-field index. An example of a single-field and key index is the index on `ObjectId` (`_id`), which is generated by default in every MongoDB collection. The `ObjectId` index is also unique, preventing a second document from having the same `ObjectId` index in a collection.

An index on a single field, based on the `mongo_book` database that we used throughout the previous chapters, is defined like this:

```
> db.books.createIndex( { price: 1 } )
```

Here, we create an index on the field name in ascending order of index creation. For descending order, the same index would be created like this:

```
> db.books.createIndex( { price: -1 } )
```

The ordering for index creation is important if we expect our queries to favor values on the first documents stored in our index. However, due to the extremely efficient time complexity that indexes have, this will not be a consideration for the most common use cases.

An index can be used for exact match queries or range queries on a field value. In the former case, the search can stop as soon as our pointer reaches the value after *O(log n)* time.

In range queries, due to the fact that we are storing values in the order in our B-tree index, once we find the border value of our range query in a node of our B-tree, we will know that all of the values in its children will be part of our result set, allowing us to conclude our search.

An example of this is shown as follows:

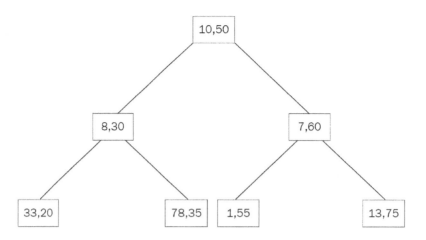

Figure 8.3 – A B-tree with data nodes

Dropping indexes

Dropping an index is as simple as creating it. We can reference the index by its name or by the fields it is composed of:

```
> db.books.dropIndex( { price: -1 } )
```

```
> db.books.dropIndex( "price_index" )
```

Indexing embedded fields

As a document database, MongoDB supports embedding fields and whole documents in nested complex hierarchies inside of the same document. Naturally, it also allows us to index these fields.

In our books collection example, we can have documents such as the following:

```
{
"_id" : ObjectId("5969ccb614ae9238fe76d7f1"),
"name" : "MongoDB Indexing Cookbook",
"isbn" : "1001",
"available" : 999,
"meta_data" : {
"page_count" : 256,
"average_customer_review" : 4.8
```

```
    }
  }
```

Here, the `meta_data` field is a document itself, with `page_count` and `average_customer_review` fields. Again, we can create an index on `page_count`, as follows:

```
db.books.createIndex( { "meta_data.page_count": 1 } )
```

This can answer queries on equality and range comparisons around the `meta_data.page_count` field, as follows:

```
> db.books.find({"meta_data.page_count": { $gte: 200 } })
> db.books.find({"meta_data.page_count": 256 })
```

> **Note**
>
> To access embedded fields, we use dot notation, and we need to include quotes (" ") around the field's name.

Indexing embedded documents

We can also index the embedded document as a whole, similar to indexing embedded fields:

```
> db.books.createIndex( { "meta_data": 1 } )
```

Here, we are indexing the whole document, expecting queries against its entirety, as in the following:

```
> db.books.find({"meta_data": {"page_count":256, "average_customer_review":4.8}})
```

The key difference is that when we index embedded fields, we can perform range queries on them using the index, whereas when we index embedded documents, we can only perform comparison queries using the index.

> **Note**
>
> The `db.books.find({"meta_data.average_customer_review": { $gte: 4.8}, "meta_data.page_count": { $gte: 200 } })` command will not use our `meta_data` index, whereas `db.books.find({"meta_data": {"page_count":256, "average_customer_review":4.8}})` will use it.

Compound indexes

Compound indexes are a generalization of single-key indexes, allowing for multiple fields to be included in the same index. They are useful when we expect our queries to span multiple fields in our documents, and also for consolidating our indexes when we start to have too many of them in our collection.

> **Note**
>
> Compound indexes can have as many as 32 fields. They can only have up to one hashed index field.

A compound index is declared similarly to single indexes by defining the fields that we want to index and the order of indexing:

```
> db.books.createIndex({"name": 1, "isbn": 1})
```

Sorting with compound indexes

The order of indexing is useful for sorting results. In single-field indexes, MongoDB can traverse the index both ways, so it doesn't matter which order we define.

In multi-field indexes, however, the ordering can determine whether we can use this index to sort. In the preceding example, a query matching the `sort` direction of our index creation will use our index as follows:

```
> db.books.find().sort( { "name": 1, "isbn": 1 })
```

It will also use a `sort` query with all of the `sort` fields reversed:

```
> db.books.find().sort( { "name": -1, "isbn": -1 })
```

In this query, since we negated both of the fields, MongoDB can use the same index, traversing it from the end to the start.

The other two sorting orders are as follows:

```
> db.books.find().sort( { "name": -1, "isbn": 1 })
> db.books.find().sort( { "name": 1, "isbn": -1 })
```

They cannot be traversed using the index, as the `sort` order that we want is not present in our index's B-tree data structure.

Reusing compound indexes

An important attribute of compound indexes is that they can be used for multiple queries on prefixes of the fields indexed. This is useful when we want to consolidate indexes that pile up in our collections over time.

Consider the compound (multi-field) index, which we created previously:

```
> db.books.createIndex({name: 1, isbn: 1})
```

This can be used for queries on `name` or `{name, isbn}`:

```
> db.books.find({"name":"MongoDB Indexing"})
> db.books.find({"isbn": "1001", "name":"MongoDB Indexing"})
```

The order of the fields in our query doesn't matter; MongoDB will rearrange the fields to match our query.

However, the order of the fields in our index does matter. A query just for the `isbn` field cannot use our index:

```
> db.books.find({"isbn": "1001"})
```

The underlying reason is that the values of our fields are stored in the index as secondary, tertiary, and so on; each one is embedded inside the previous ones, just like a **matryoshka**, the Russian nesting doll. This means that when we query on the first field of our multi-field index, we can use the outermost doll to find our pattern, whereas when we are searching for the first two fields, we can match the pattern on the outermost doll and then dive into the inner one.

This concept is called **prefix indexing**, and along with index intersection, it is the most powerful tool for index consolidation, as you will see later in this chapter.

Multikey indexes

Indexing scalar (single) values was explained in the preceding sections. One of the advantages that we have when using MongoDB is the ability to easily store vector values in the form of arrays.

In the relational world, storing arrays is generally frowned upon, as it violates the normal forms of normalization. In a document-oriented database such as MongoDB, it is frequently a part of our design, as we can store and easily query complex structures of data.

Indexing arrays of documents is achieved by using the multikey index. A multikey index can store both arrays of scalar values and arrays of nested documents.

Creating a multikey index is the same as creating a regular index:

```
> db.books.createIndex({"tags":1})
```

Assume that we have created a document in our books collection, using the following command:

```
> db.books.insert({"name": "MongoDB Multikeys Cheatsheet",
"isbn": "1002", "available": 1, "meta_data": {"page_
count":128, "average_customer_review":3.9}, "tags": ["mongodb",
"index","cheatsheet","new"] })
```

Our new index will be a multikey index, allowing us to find documents with any of the tags stored in our array:

```
> db.books.find({tags:"new"})
{
"_id" : ObjectId("5969f4bc14ae9238fe76d7f2"),
"name" : "MongoDB Multikeys Cheatsheet",
"isbn" : "1002",
"available" : 1,
"meta_data" : {
"page_count" : 128,
"average_customer_review" : 3.9
},
"tags" : [
"mongodb",
"index",
"cheatsheet",
"new"
]
}
>
```

We can also create compound indexes with a multikey index, but we can only have one array in each and every index document at the most. Given that in MongoDB we don't specify the type of each field, this means that creating an index with two or more fields with an array value will fail at creation time, and trying to insert a document with two or more fields as arrays will fail at insertion time.

For example, an analytics_data compound index on tags will fail to be created if we have the following document in our database:

```
{
"_id" : ObjectId("5969f71314ae9238fe76d7f3"),
"name": "Mastering parallel arrays indexing",
```

```
"tags" : [
"A",
"B"
],
"analytics_data" : [
"1001",
"1002"
]
}


> db.books.createIndex({tags:1, analytics_data:1})
{
"ok" : 0,
"errmsg" : "cannot index parallel arrays [analytics_data]
[tags]",
"code" : 171,
"codeName" : "CannotIndexParallelArrays"
}
```

Consequently, if we first create the index on an empty collection and try to insert this document, the insert will fail with the following error:

```
> db.books.find({isbn:"1001"}).hint("international_standard_
book_number_index").explain()
{
  "queryPlanner" : {
  "plannerVersion" : 1,
  "namespace" : "mongo_book.books",
  "indexFilterSet" : false,
  "parsedQuery" : {
  "isbn" : {
  "$eq" : "1001"
  }
  },
  "winningPlan" : {
  "stage" : "FETCH",
  "inputStage" : {
  "stage" : "IXSCAN",
```

```
 "keyPattern" : {
 "isbn" : 1
 },
 "indexName" : "international_standard_book_numbe
r_index",
 "isMultiKey" : false,
 "multiKeyPaths" : {
 "isbn" : [ ]
 },
 "isUnique" : false,
 "isPartial" : false,
 "indexVersion" : 2,
 "direction" : "forward",
 "indexBounds" : {
 "isbn" : [
 "[\"1001\", \"1001\"]"
 ]
 }
 }
 },
 "rejectedPlans" : [ ]
 },
 "serverInfo" : {
 "host" : "PPMUMCPU0142",
 "port" : 27017,
 "version" : "3.4.7",
 "gitVersion" : "cf38c1b8a0a8dca4a11737581beafef4fe120bcd"
 },
 "ok" : 1
```

> **Note**
> Hashed indexes cannot be multikey indexes.

Another limitation that we will likely run into when trying to fine-tune our database is that multikey indexes cannot cover a query completely. A compound index with multikey fields can be used by the query planner only on the non-multikey fields.

Covering a query with the index means that we can get our result data entirely from the index, without accessing the data in our database at all. This can result in dramatically increased performance, as indexes are most likely to be stored in RAM.

Querying for multiple values in multikey indexes will result in a two-step process from the index's perspective.

In the first step, the index will be used to retrieve the first value of the array, and then a sequential scan will run through the rest of the elements in the array; an example is as follows:

```
> db.books.find({tags: [ "mongodb", "index", "cheatsheet",
"new" ] })
```

This will first search for all entries in multikey index tags that have a mongodb value and will then sequentially scan through them to find the ones that also have the index, cheatsheet, and new tags.

> **Note**
>
> A multikey index cannot be used as a shard key. However, if the shard key is a prefix index of a multikey index, it can be used. We will cover more on this in *Chapter 14, Mastering Sharding*.

Special types of indexes

Besides the generic indexes, MongoDB supports indexes for special use cases. In this section, we will identify and explore how to use them.

Text indexes

Text indexes are special indexes on string value fields, which are used to support text searches. This book is based on version 3 of the text index functionality, available since version 3.2.

A text index can be specified similarly to a regular index by replacing the index sort order (-1, 1) with text, as follows:

```
> db.books.createIndex({"name": "text"})
```

> **Note**
>
> A collection can have one text index at most. This text index can support multiple fields, whether text or not. It cannot support other special types, such as multikey or geospatial. Text indexes cannot be used for sorting results, even if they are only a part of a compound index.
>
> **MongoDB Atlas Search** allows for multiple full-text search indexes. MongoDB Atlas Search is built on **Apache Lucene** and thus is not limited by MongoDB on-premise full-text search.

Since we only have one text index per collection, we need to choose the fields wisely. Reconstructing this text index can take quite some time and having only one per collection makes maintenance quite tricky, as you will see toward the end of this chapter.

We can use the same syntax as with a wildcard index and index each and every field in a document that contains strings as a `text` index:

```
> db.books.createIndex( { "$**": "text" } )
```

This can result in unbounded index sizes and should be avoided; however, it can be useful if we have unstructured data (for example, coming straight from application logs wherein we don't know which fields may be useful, and we want to be able to query as many of them as possible).

A compound index can also be a `text` index:

```
> db.books.createIndex( { "available": 1, "meta_data.page_
count": 1, "$**": "text" } )
```

A compound index with `text` fields follows the same rules of sorting and prefix indexing that we explained earlier in this chapter. We can use this index to query on `available`, or the combination of `available` and `meta_data.page_count`, or sort them if the sort order allows for traversing our index in any direction.

Text indexes will apply stemming (removing common suffixes, such as plural `s`/`es` for English language words) and remove stop words (`a`, `an`, `the`, and so on) from the index. Text indexes behave the same as partial indexes, not adding a document to the text index if it lacks a value for the field that we have indexed it on.

We can control the weights of different text-indexed fields, such as the following:

```
> db.books.createIndex({
"book_name": "text",
"book_description": "text" },
{
    weights: {
    book_name: 10,
    book_description: 5
    },
    name: "TextIndex"
})
```

In the preceding sample code, assuming a document with book_name and book_description, we have assigned a weight of 10 for book_name and a weight of 5 for book_description.

Search matches in `book_name` will be scored twice as much (10/5 = 2) as matches in `book_description`, affecting the full-text search results.

> **Note**
>
> Text indexing supports 15 languages as of MongoDB 5.3, including Spanish, Portuguese, and German. Text indexes require special configurations to correctly index in languages other than English.

Some interesting properties of text indexes are explained as follows:

- **Case-insensitivity and diacritic insensitivity**: A text index is case-insensitive and diacritic-insensitive. Version 3 of the text index (the one that comes with version 3.4) supports common *C*, simple *S*, and the special *T* case folding, as described in the **Unicode Character Database (UCD)** 8.0 case folding. In addition to case insensitivity, version 3 of the text index supports diacritic insensitivity. This expands insensitivity to characters with accents in both small and capital letter forms. For example, *e*, *è*, *é*, *ê*, *ë*, and their capital letter counterparts, can all be equal when compared using a text index. In the previous versions of the text index, these were treated as different strings.

- **Tokenization delimiters**: Version 3 of the text index supports tokenization delimiters, defined as `Dash`, `Hyphen`, `Pattern_Syntax`, `Quotation_Mark`, `Terminal_Punctuation`, and `White_Space`, as described in the UCD 8.0 case folding.

Full-text search in MongoDB is considered a legacy feature and MongoDB Atlas Search is way more powerful and flexible.

Hashed indexes

A hashed index contains `hashed` values of the indexed field:

```
> db.books.createIndex( { name: "hashed" } )
```

This will create a hashed index on the name of each book in our `books` collection. A hashed index is ideal for equality matches but it cannot work with range queries. If we want to perform a range of queries on fields, we can create a compound index with at most one hashed field, which will be used for equality matches.

The following mongo shell code will create a compound index on `created_at_date` and `name`, with the `name` field being the one and only hashed field that we can create in a compound index:

```
> db.books.createIndex( { created_at_date: 1, name: "hashed" }
)
```

We cannot create a hashed index on an array field.

We cannot enforce uniqueness on a hashed field value. We can add a single unique index on the same field separately instead.

Hashed indexes are used internally by MongoDB for hash-based sharding, as we will discuss in *Chapter 14, Mastering Sharding*. Hashed indexes truncate floating point fields to integers. Floating points should be avoided for hashed fields wherever possible.

Time-to-live indexes

Time-to-live (TTL) indexes are used to automatically delete documents after an expiration time. Their syntax is as follows:

```
> db.books.createIndex( { created_at_date: 1 }, {
expireAfterSeconds: 86400 } )
```

The `created_at_date` field values have to be either a date or an array of dates (the earliest one will be used). In this example, the documents will be deleted one day (`86400` seconds) after `created_at_date`.

If the field does not exist or the value is not a date, the document will not expire. In other words, a TTL index silently fails and does not return any errors when it does.

Data gets removed by a background job, which runs every 60 seconds. As a result, there is no explicitly guaranteed accurate measure of how much longer documents will persist past their expiration dates.

> **Note**
>
> A TTL index is a regular single-field index. It can be used for queries like a regular index. A TTL index cannot be a compound index, operate on a capped or time series collection, or use the `_id` field. The `_id` field implicitly contains a timestamp of the created time for the document but is not a `Date` field. If we want each document to expire at a different, custom date point, we have to set `{expireAfterSeconds: 0}`, and set the TTL index `Date` field manually to the date on which we want the document to expire.

Partial indexes

A partial index on a collection is an index that only applies to the documents that satisfy the `partialFilterExpression` query.

We'll use our familiar `books` collection, as follows:

```
> db.books.createIndex(
  { price: 1, name: 1 },
  { partialFilterExpression: { price: { $gt: 30 } } }
)
```

Using this, we can have an index only for books that have a price greater than 30. The advantage of partial indexes is that they are more lightweight in creation and maintenance and use less storage than an index on every possible value.

The `partialFilterExpression` filter supports the following operators:

- Equality expressions (that is, `field: value`, or using the `$eq` operator)

- The `$exists: true` expression

- The `$gt`, `$gte`, `$lt`, and `$lte` expressions

- `$type` expressions

- The `$and` operator, at the top level only

Partial indexes will only be used if the query can be satisfied as a whole by the partial index.

If our query matches or is more restrictive than the `partialFilterExpression` filter, then the partial index will be used. If the results may not be contained in the partial index, then the index will be totally ignored.

`partialFilterExpression` does not need to be a part of the sparse index fields. The following index is a valid sparse index:

```
> db.books.createIndex({ name: 1 },{ partialFilterExpression:
{ price: { $gt: 30 } } })
```

To use this partial index, however, we need to query for both `name` and `price` equal to or greater than 30.

Earlier versions of MongoDB offered a subset of the partial indexes functionality, named sparse indexes. Partial indexes were introduced in MongoDB 3.2 and are recommended over the more restrictive sparse indexes, so if you have sparse indexes from earlier versions, it may be a good idea to upgrade them. The `_id` field cannot be a part of a partial index.

A shard key index or an `_id` index cannot be a partial index.

We can use the same key pattern (for example, `db.books.createIndex( {name: 1, price: 1} )`) to create multiple partial indexes on the same collection but the `partialFilterExpression` filter has to be distinct between them.

A unique index can also be a partial index at the same time and MongoDB will enforce the uniqueness for documents that satisfy the `partialFilterExpression` filter. Documents that do not satisfy the `partialFilterExpression` filter, that have null or the field does not exist at all, will not be subject to the unique constraint.

Unique indexes

A unique index is similar to an RDBMS unique index, forbidding duplicate values for the indexed field. MongoDB creates a unique index by default on the _id field for every inserted document:

```
> db.books.createIndex( { "name": 1 }, { unique: true } )
```

This will create a unique index on a book's name. A unique index can also be a compound embedded field or an embedded document index.

In a compound index, the uniqueness is enforced across the combination of values in all of the fields of the index; for example, the following will not violate the unique index:

```
> db.books.createIndex( { "name": 1, "isbn": 1 }, { unique:
true } )
> db.books.insert({"name": "Mastering MongoDB", "isbn": "101"})
> db.books.insert({"name": "Mastering MongoDB", "isbn": "102"})
```

This is because even though the name is the same, our index is looking for the unique combination of name and isbn, and the two entries differ in isbn.

Unique indexes do not work with hashed indexes. Unique indexes cannot be created if the collection already contains duplicate values of the indexed field. A unique index will not prevent the same document from having multiple values.

The only way that we can enforce a unique index across different shards is if the shard key is a prefix or the same as the unique compound index. Uniqueness is enforced in all of the fields of the unique compound index.

If a document is missing the indexed field, it will be inserted. If a second document is missing the indexed field, it will not be inserted. This is because MongoDB will store the missing field value as null, only allowing one document to be missing in the field.

Indexes that are a combination of unique and partial will only apply unique indexes after a partial index has been applied. This means that there may be several documents with duplicate values if they are not a part of partial filtering.

Case insensitivity

Case sensitivity is a common problem with indexing. We may store our data in mixed caps and need our index to ignore cases when looking for our stored data. Until version 3.4, this was dealt with at the application level by creating duplicate fields with all lowercase characters and indexing all lowercase fields to simulate a case-insensitive index.

Using the collation parameter, we can create case-insensitive indexes and even collections that behave as case-insensitive.

In general, `collation` allows users to specify language-specific rules for string comparisons. A possible (but not the only) usage is for case-insensitive indexes and queries.

Using our familiar `books` collection, we can create a case-insensitive index on a name, as follows:

```
> db.books.createIndex( { "name" : 1 },
  { collation: {
  locale : 'en',
  strength : 1
  }
  } )
```

The `strength` parameter is one of the collation parameters, the defining parameter for case-sensitivity comparisons. Strength levels follow the **International Components for Unicode (ICU)** comparison levels. The values that it accepts are as follows:

Strength Value	Description
1	The primary level of comparison. Comparison based on string values, ignoring any other differences, such as case and diacritics.
2	The secondary level of comparison is the comparison based on the primary level, and if this is equal, then it compares diacritics (that is, accents).
3 (default)	The tertiary level of comparison. As with level 2, adding case and variants.
4	The quaternary level. Limited for specific use cases to consider the punctuation when levels 1-3 ignore punctuation, or for processing Japanese text.
5	Identical level. Limited for specific use cases: a tie-breaker.

Table 8.1 – Collation strength values

Creating the index with `collation` is not enough to get back case-insensitive results. We need to specify `collation` in our query as well:

```
> db.books.find( { name: "Mastering MongoDB" } ).collation( {
  locale: 'en', strength: 1 } )
```

If we specify the same level of `collation` in our query as our index, then the index will be used. We could specify a different level of `collation`, as follows:

```
> db.books.find( { name: "Mastering MongoDB" } ).collation( {
  locale: 'en', strength: 2 } )
```

Here, we cannot use the index, as our index has a level 1 `collation` parameter, and our query looks for a level 2 `collation` parameter.

> **Note**
>
> Diacritics, also known as diacritical marks, are one or more characters that have a mark near, above, or through them to indicate a different phonetic value than the unmarked equivalents. A common example is the French é, as in café.

If we don't use any collation in our queries, we will get results defaulting to level 3, that is, case-sensitive.

Indexes in collections that were created using a different `collation` parameter from the default will automatically inherit this `collation` level.

Suppose that we create a collection with a level `1` `collation` parameter, as follows:

```
> db.createCollection("case_sensitive_books", { collation: {
locale: 'en_US', strength: 1 } } )
```

The following index will also have `strength: 1` collation because the index inherits the collation strength from the collection:

```
> db.case_sensitive_books.createIndex( { name: 1 } )
```

Default queries to this collection will be `strength: 1` collation, case-sensitive, and ignoring diacritics. If we want to override this in our queries, we need to specify a different level of collation in our queries or ignore the `strength` part altogether. The following two queries will return case-insensitive, default `collation` level results in our `case_sensitive_books` collection:

```
> db.case_sensitive_books.find( { name: "Mastering MongoDB" }
).collation( { locale: 'en', strength: 3 } ) // querying the
colletion with the global default collation strength value, 3
> db.case_sensitive_books.find( { name: "Mastering MongoDB"
} ).collation( { locale: 'en' } ) // no value for collation,
will reset to global default (3) instead of default for case_
sensitive_books collection (1)
```

Collation is a very useful concept in MongoDB and we will continue exploring it throughout the following chapters.

Geospatial indexes

Geospatial indexes were introduced early on in MongoDB and the fact that Foursquare was one of the earliest customers and success stories for MongoDB (then 10gen Inc.) is probably no coincidence. There are three distinct types of geospatial indexes that we will explore in this chapter and they will be covered in the following sections.

2D geospatial indexes

A 2D geospatial index stores geospatial data as points on a two-dimensional plane. It is mostly kept for legacy reasons, for coordinate pairs created before MongoDB 2.2, and in most cases, it should not be used with the latest versions.

2dsphere geospatial indexes

A 2dsphere geospatial index supports queries calculating geometries on an Earth-like plane. It is more accurate than the simplistic 2D index and can support both GeoJSON objects and legacy coordinate pairs as input. It is the recommended type of index to use for geospatial queries.

Its current version, since MongoDB 3.2, is version 3. It is a sparse index by default, only indexing documents that have a 2dsphere field value. Assuming that we have a location field in our books collection, tracking the home address of the main author of each book, we could create an index on this field as follows:

```
> db.books.createIndex( { location: "2dsphere" } )
```

The location field needs to be a GeoJSON object, such as this one:

```
location : { type: "Point", coordinates: [ 51.5876, 0.1643 ] }
```

A 2dsphere index can also be a part of a compound index, in any position in the index, first or not:

```
> db.books.createIndex( { name: 1, location : "2dsphere" })
```

Wildcard indexes

MongoDB is **schemaless**. As such, there are cases where we may not exactly know the names of the fields beforehand.

Wildcard indexes allow us to define patterns for the names of the fields that MongoDB will index (include) or not index (exclude). We generally cannot mix these two conditions at the collection level, as this could result in conflicting rules, which would require manual intervention.

> **Note**
> The only case where we can mix the inclusion and exclusion of fields is the _id field. We can create a wildcard index with _id:0 or _id:1 but all the other fields in the wildcard declaration have to be either entirely included or entirely excluded.

To create a wildcard index in all attributes of a collection, we have to use the following command:

```
> db.books.createIndex( { "$**": 1 } )
```

> **Note**
>
> The catchall wildcard index on every attribute of the collection will recursively create indexes for subdocuments and any attributes of the array type.

We can also create wildcard indexes in subdocuments.

Given a collection, books with a document will look like this:

```
{ name: "Mastering MongoDB 5.X", attributes: { author: "Alex
Giamas", publisher: "Packt" }, chapters: [{id: 1, name:
"MongoDB - A database for modern web" }] }
```

We can index the document with embedded attributes:

```
> db.books.createIndex( { "attributes.$**": 1 } )
```

The only difference from the generic wildcard index creation is that we are now targeting the attributes subdocument with the $** wildcard operator.

We can also create wildcard indexes on specific attributes of the document, both for inclusion and exclusion. To achieve this, we need to use the `wildcardProjection` operator, which behaves similarly to the `$project` phase that we discussed previously, in *Chapter 7, Aggregation*.

An example of creating a wildcard index with inclusion follows:

```
> db.books.createIndex( { "$**": 1 }, {
"wildcardProjection": {
{ "name": 1, "attributes.author": 1 }
}
} )
```

This will create a wildcard index on the name attribute, the `attributes.author` attribute, and any subdocuments and arrays that the fields may contain.

This index will not index the chapter's attribute and any subdocuments in its array.

Similarly, if we want to create a wildcard index with exclusion, we can do it as follows:

```
> db.books.createIndex( { "$**": 1 }, {
"wildcardProjection": {
{ "chapters": 0 }
}
} )
```

This index will index all attributes and their subdocuments, except for the `chapters` field.

For the sample document mentioned above, the last two indexes are equivalent.

> **Note**
>
> Wildcard indexes are not a replacement for proper index design. Having a large and unbounded number of indexes in a collection will result in performance degradation, no matter how we generated the indexes.

The most important limitations of wildcard indexes are listed here:

- Wildcard index generation creates single attribute indexes across all fields. This means that they are mostly useful in single-field queries and will use at most one field to optimize the query.
- Wildcard indexes are essentially partial indexes and as such, they cannot support a query looking for attributes that do not exist in the documents.
- Wildcard indexes will recursively index every attribute of the subdocuments and as such, cannot be used to query for exact subdocument or whole array equality ($eq) or inequality ($neq) queries.
- Wildcard indexes cannot be used as keys for sharding purposes.

Hidden indexes

Simply put, a hidden index is an index that exists but is not visible to the query planner. A hidden index will not be used to speed up any query.

A hidden index still exists; it's just not used. As such, if it is a unique index, for example, it will still apply the unique constraint. Likewise, if it is a TTL index, it will keep expiring documents.

We cannot hide the _id special index.

To add a hidden index, we need to pass the hidden: true parameter, as shown here:

```
> db.books.createIndex( { name: 1 }, { hidden: true, name:
"index_on_name_field" } )
```

To unhide an existing index, we can use unhideIndex, either with the index name or the fields that the index uses.

For example, to unhide the index that we just created, we could use either of the following commands respectively:

```
> db.books.unhideIndex("index_on_name_field")
> db.books.unhideIndex({name: 1})
```

To hide an existing index, we can use `hideIndex`, either with the index name or with the fields that the index uses.

For example, to hide the index that we just unhid again, we could use either of the following commands respectively:

```
> db.books.hideIndex("index_on_name_field")
> db.books.hideIndex({name: 1})
```

Clustered indexes

A new addition since MongoDB 5.3, clustered indexes allow us to perform faster queries on the clustered index key value. The documents are physically stored in order of the clustered key index value, meaning that we get significant performance improvement and space savings, at the expense of having to query using the clustered index.

We are free to create secondary indexes in addition to the clustered index on a clustered collection but this will negate the above-mentioned advantages.

We can define any value we want but we need to set our custom value in the `_id` key. The clustered index cannot be hidden.

A clustered collection needs to be created as such and we cannot convert an existing collection into a clustered one. The clustered collection can also be a TTL collection if we set the `expireAfterSeconds` parameter to the creation time.

A clustered collection can be more space efficient than a non-clustered (default) collection but we need to explicitly skip creating a secondary index at creation time. The query optimizer will try to use the secondary index instead of the clustered index if we don't provide a hint to skip it.

The key that we will use to populate the `_id` attribute in the collection on our own must be unique, immutable, and as small in size as possible and it's strongly advisable to be monotonically increasing in value.

A clustered collection can save storage space, especially when we skip creating the secondary index, and can serve our queries strictly by using the clustered index. The clustered index was just introduced in version 6.0 and we expect it to be improved in upcoming versions.

In this section, we learned about all the different types of special indexes that MongoDB offers. In the next section, we will understand how we can build and manage our indexes.

Building and managing indexes

Indexes can be built using the MongoDB shell or any of the available drivers.

MongoDB was used to build indexes in the foreground (using faster, blocking operations) or the background (using slower, non-blocking operations) up until version 4.4.

From this version onward, all indexes are now built using an optimized build process, which is at least as fast as the slower background build option present in previous versions.

The fewer inserts and updates that occur during the build process, the faster the index creation will be. Few to no updates will result in the build process happening as fast as the former foreground build option.

Index builds now obtain an exclusive lock at the collection level only for a short period at the beginning and the end of the build process. The rest of the time, the index build yields for read-write access.

An index constraint violation occurring from documents pre-existing to the build index command or being added during index creation will only fail the index creation at the end of the index build process.

Ensure that you index early and consolidate indexes by revisiting them regularly. Queries won't see partial index results. Queries will start getting results from an index only after it is completely created.

Do not use the main application code to create indexes, as it can impose unpredictable delays. Instead, get a list of indexes from the application, and mark these for creation during maintenance windows.

Index filters

An index filter can force the MongoDB query plan to consider a specific set of indexes for a specific query shape.

A query shape is the combination of the fields that we are querying, along with any sorting, projection, and collation options that we specify in the query.

For example, in the following query, this is what we do:

```
> db.books.find({isbn: "1001"})
```

The query shape only consists of the isbn field name. The value that we are querying is not important.

To declare an index filter on the isbn field for our books collection, we can use the following command:

```
> db.runCommand({
planCacheSetFilter: "isbn_lookup",
query: { isbn: "any_string_here" },
indexes: [
    { isbn: 1 },
    "isbn_book_name_index",
]})
```

With the preceding command, the query planner will consider the index on { isbn: 1 }, the isbn_book_name_index index, and a full collection scan, which is always considered as a last resort by the query planner.

An index filter is more powerful than a hint() parameter. MongoDB will ignore any hint() parameter if it can match the query shape to one or more index filters.

Index filters are only held in memory and will be erased on server restart.

Forcing index usage

We can force MongoDB to use an index by applying the hint() parameter:

```
> db.books.createIndex( { isbn: 1 } )
{
"createdCollectionAutomatically" : false,
"numIndexesBefore" : 8,
"numIndexesAfter" : 9,
"ok" : 1
}
```

The output from createIndex notifies us that the index was created ("ok" : 1), no collection was automatically created as a part of index creation ("createdCollectionAutomatically" : false), the number of indexes before this index creation was 8, and now there are nine indexes for this collection in total.

Now, if we try to search for a book by isbn, we can use the explain() command to see the winningPlan subdocument, where we can find which query plan was used:

```
> db.books.find({isbn: "1001"}).explain()
...
"winningPlan" : {
"stage" : "FETCH",
"inputStage" : {
"stage" : "IXSCAN",
"keyPattern" : {
"isbn" : 1,
"name" : 1
},
"indexName" : "isbn_1_name_1",
. . .
```

This means that an index with an isbn field of 1 and a name field of 1 was used instead of our newly created index. We can also view our index in the rejectedPlans subdocument of the output, as follows:

```
...
"rejectedPlans" : [
{
"stage" : "FETCH",
"inputStage" : {
"stage" : "IXSCAN",
"keyPattern" : {
"isbn" : 1
},
"indexName" : "isbn_1",
. . .
```

This is, in fact, correct, as MongoDB is trying to reuse an index that is more specific than a generic one.

We may not be sure though in cases where our isbn_1 index is performing better than the isbn_1_name_1 index.

We can force MongoDB to use our newly created index, as follows:

```
> db.books.find({isbn: "1001"}).hint("international_standard_
book_number_index")
.explain()
{
. . .
 "winningPlan" : {
 "stage" : "FETCH",
 "inputStage" : {
 "stage" : "IXSCAN",
 "keyPattern" : {
 "isbn" : 1
 },
. . .
```

Now, the winningPlan subdocument contains our isbn_1 index, and there are no items in the rejectedPlans subdocument.

We can also force the index that the query will use by its definition, such that the next two commands are equivalent:

```
> db.books.find({isbn: "1001"}).hint("international_standard_
book_number_index")
> db.books.find({isbn: "1001"}).hint( { isbn: 1 } )
```

In this example, we are declaring the index to force MongoDB to use its name and its declaration. The two commands are equivalent.

> **Note**
>
> We cannot use `hint()` when the query contains a `$text` query expression or while an index is hidden.
>
> The MongoDB query planner will always choose the index filter and ignore the `hint()` parameter if an index filter already exists for our query.

We can force the query planner to skip any index by using the special `{ $natural: <traversal order> }` index declaration. `<traversal order>` can be 1 or -1 for forward scanning or backward scanning of all the documents in the collection.

Disabling the index lookup in a query is not recommended in general but may help us debug edge cases when query results don't match what we are expecting.

Rolling index building on replica sets and sharded environments

In replica sets, if we issue a `createIndex()` command, the operation will start simultaneously in all data-bearing replica set members.

Similarly, in a sharded environment, all the data-bearing members across the replica sets that contain data for our collection will simultaneously start building the index.

The recommended approach to building indexes in replica sets when we want to avoid any performance degradation goes as follows:

- Stop one secondary from the replica set
- Restart it as a standalone server in a different port
- Build the index from the shell as a standalone index
- Restart the secondary in the replica set
- Allow for the secondary to catch up with the primary

We need to have a large enough oplog size in the primary to make sure that the secondary will be able to catch up once it's reconnected. The oplog size is defined in MB in the configuration and it defines how many operations will be kept in the log in the primary server. If the oplog size can only hold up to the last 100 operations happening in the primary, and 101 or more operations happen, this means that the secondary will not be able to sync with the primary. This is a consequence of the primary not having enough memory to keep track of its operations and inform the secondary about them.

This approach can be repeated for each secondary server in the replica set. Finally, we can repeat the same procedure for the primary server. First, we step down the primary server using `rs.stepDown()`, restart it as a standalone server, build the index, and add it back into the replica set.

The `rs.stepDown()` command will not step down the server immediately but will instead wait for a default of 10 seconds (configurable using the `secondaryCatchUpPeriodSecs` parameter) for any eligible secondary server to catch up to its data.

For a sharded environment, we have to stop the sharding data balancer first. Then, we need to find out which shards (here, meaning the replica sets) contain the collection(s) that we need to index. Finally, we need to repeat the process for each shard/replica set, as explained already.

Since version 4.2, MongoDB builds indexes using an optimized build process, which doesn't lock the database for extended periods, and as such, rolling index builds are less frequently necessary after this version.

That being said, building indexes can be risky and operationally intensive. It is always a good idea to have a staging environment that mirrors production and to dry run operations that affect the live cluster in staging in order to avoid surprises.

Managing indexes

In this section, you will learn how to give human-friendly names to your indexes, as well as some special considerations and limitations that we have to keep in mind when indexing.

Naming indexes

By default, index names are assigned by MongoDB automatically based on the fields indexed and the direction of the index (1 or -1). If we want to assign our own `name`, we can do so at creation time:

```
> db.books.createIndex( { isbn: 1 }, { name: "international_
standard_book_number_index" } )
```

Now, we have a new index called `international_standard_book_number_index`, instead of what MongoDB would have named it (`"isbn_1"`).

> **Note**
>
> We can view all of the indexes in our `books` collection by using `db.books.getIndexes()`.

Special considerations

The following are a few limitations to keep in mind concerning indexing:

- A collection can have up to 64 indexes.

- A compound index can have up to 32 fields.

- Geospatial indexes cannot cover a query. Multikey indexes cannot cover a query over fields of an array type. This means that index data alone will not be enough to fulfill the query and the underlying documents will need to be processed by MongoDB to get back a complete set of results.

- Indexes have a unique constraint on fields. We cannot create multiple indexes on the same fields, differing only in options. We can, however, create multiple partial indexes with the same key pattern given that the filters in `partialFilterExpression` differ.

Using indexes efficiently

Creating indexes is a decision that should not be taken lightly. As easy as it is to create indexes via the shell, it can create problems down the line if we end up with too many or with inadequately efficient indexes. In this section, we will learn how to measure the performance of existing indexes, some tips for improving performance, and how we can consolidate the number of indexes so that we have better-performing indexes.

Measuring performance

Learning how to interpret the `explain()` command will help you with both optimizing and understanding the performance of an index. The `explain()` command, when used in conjunction with a query, will return the query plan that MongoDB would use for this query, instead of the actual results.

It is invoked by chaining it at the end of our query, as follows:

```
> db.books.find().explain()
```

It can take three options: `queryPlanner` (the default), `executionStats`, and `allPlansExecution`.

Let's use the most verbose output, `allPlansExecution`:

```
> db.books.find().explain("allPlansExecution")
```

Here, we can get information for both the winning query plan and some partial information about query plans, which were considered during the planning phase but were rejected because the query planner considered them slower. The `explain()` command returns a rather verbose output in all cases, allowing for deeper insight into how the query plan works to return our results.

MongoDB version 5.1 introduced a new query execution engine, the **slot-based query execution engine (SBE)**. The slot-based execution engine is preferable for the query planner where possible, else it falls back to the older query planner. When the query planner is using the new engine, we will see the output in `explain()` like so:

```
{ ... explainVersion: 2 ... }
```

`explainVersion: 1` means that the query planner is using the old engine.

The query plans are shown as a tree of stages, with each step of each subsequent step embedded in the previous step's document. We can also use the `planNodeId` integer field in the output to identify the different query execution steps. More low-level details of each stage of SBE are also available in the `slotBasedPlan` field. These are pretty low-level and are mostly used for debugging by MongoDB engineers when there are unexpected results.

At first glance, we need to focus on whether the indexes that should be used are being used, and whether the number of scanned documents matches the number of returned documents as closely as possible.

For the first one, we can inspect the `stage` field and look for `IXSCAN`, which means that an index was used. Then, in the sibling `indexName` field, we are able to see the name of our expected index.

For the second one, we need to compare `keysExamined` with the `nReturned` fields for the old execution engine. We ideally want our indexes to be as selective as possible with regard to our queries, meaning that to return 100 documents, these would be the 100 documents that our index examines.

The new execution engine (SBE) will include these fields in the `$planCacheStats` for each collection. The `creationExecStats` subdocument array will include one entry for each of the cached execution plans.

The number of keys examined is `totalKeysExamined` and the number of documents examined is named `totalDocsExamined`, whereas the `nReturned` field shows the number of documents returned.

The MongoDB SBE will calculate a unique `queryHash` for each query shape. A query shape is the combination of the fields that we are querying together with sorting, projection, and collation.

We can list the query plans that are cached for each collection, flush the cache, or increase its size. Subsequent queries using the same query shape will use the cached plan instead of calculating a new one. The cached plans are stored in memory, so will be deleted on server shutdown.

Finally, there is a trade-off as indexes increase in number and size in our collection. We can have a limited number of indexes per collection, and we definitely have a limited amount of RAM to fit these indexes into, so we must balance the trade-off between having the best available indexes and these indexes not fitting into our memory and getting slowed down because of disk swapping.

Improving performance

Once we get comfortable with measuring the performance of the most common and important queries for our users, we can start to try to improve them.

The general idea is that we need indexes when we expect (or already have) repeatable queries that are starting to run slowly. Indexes do not come for free, as they impose a performance penalty in document creation and maintenance, but they are more than worth it for frequent queries and can reduce the contention in our database if designed correctly.

Recapping our suggestions from the previous section, we want our indexes to do the following:

- Fit in the RAM
- Ensure selectivity
- Be used to sort our query results
- Be used in our most common and important queries

Fitting in the RAM can be ensured by using `getIndexes()` in our collections and making sure that we are not creating large indexes by inspecting the system-level available RAM and whether swapping is being used or not.

Selectivity, as mentioned previously, is ensured by comparing `nReturned` with `keysExamined` in each `IXSCAN` phase of our queries. We want these two numbers to be as similar as possible.

Ensuring that our indexes are used to sort our query results involves a combination of using compound indexes (which will be used as a whole, and also for any prefix-based query) and declaring the direction of our indexes to be in accordance with our most common queries.

Finally, aligning indexes with our query is a matter of application usage patterns, which can uncover that queries are used most of the time, and then using `explain()` on these queries to identify the query plan that is being used each time.

Index intersection

Index intersection refers to the concept of using more than one index to fulfill a query.

Index intersection can happen when we use OR (`$or`) queries by using a different index for each OR clause. Index intersection can happen when we use AND queries and we have either complete indexes for each AND clause or index prefixes for some (or all) of the clauses.

For example, consider a query on our `books` collection, as follows:

```
> db.books.find({ "isbn":"101", "price": { $gt: 20 }})
```

Here, with two indexes (one on `isbn` and the other on `price`), MongoDB can use each index to get the related results, and then intersect the index results to get the result set.

With compound indexes, as you learned previously in this chapter, we can use index prefixing to support queries that contain the first *1...n-1* fields of an *n* field compound index.

What we cannot support with compound indexes are queries that are looking for fields in the compound index, missing one or more of the previously defined fields.

> **Note**
> The order matters in compound indexes.

To satisfy these queries, we can create indexes on the individual fields, which will then use index intersection and fulfill our needs. The downside to this approach is that as the number of fields (*n*) increases, the number of indexes that we have to create grows exponentially, thus increasing our need for storage and memory.

Index intersection will not work with `sort()`. We can't use one index for querying and a different index for applying `sort()` to our results.

However, if we have an index that can fulfill a part of the whole of our query and the `sort()` field, then this index will be used. The query planner will usually avoid using sorting with index intersection. Index intersection with sorting will show up as an `AND_SORTED` stage in the executed query plan.

We should not rely on index intersection for our schema design and should always use compound indexes where possible.

Further reading

You can refer to the following links for further references:

- A cheat sheet on big O notation

  `http://bigocheatsheet.com/`

- The MongoDB documentation on index intersection

  `https://docs.mongodb.com/manual/core/index-intersection/`

Summary

In this chapter, we learned about the foundations of indexing and index internals. We then explored how to use the different index types available in MongoDB, such as single-field, compound, and multikey, as well as some special types, such as text, hashed, TTL, partial, parse, unique, case-insensitive, and geospatial.

In the next part of the chapter, we learned how to build and manage indexes using the shell, which is a basic part of administration and database management, even for NoSQL databases. Finally, we discussed how to improve our indexes at a high level, and also how we can use index intersection in practice to consolidate the number of indexes.

In the next chapter, we will discuss how we can monitor our MongoDB cluster and keep consistent backups. We will also learn how to handle security in MongoDB.

Part 3 – Administration and Data Management

This part of the book will focus on the non-functional aspects of MongoDB. We will learn how to monitor and secure our data and how to avoid data loss in the event of a catastrophic failure. We will learn about different underlying storage options and how to perform administration tasks in MongoDB clusters. Finally, MongoDB tooling will cover the MongoDB Atlas ecosystem of tools and the equivalent options we have if we decide to host our clusters on-premises or through our own cloud provider.

This part contains the following chapters:

- *Chapter 9, Monitoring, Backup, and Security*
- *Chapter 10, Managing Storage Engines*
- *Chapter 11, MongoDB Tooling*
- *Chapter 12, Harnessing Big Data with MongoDB*

9

Monitoring, Backup, and Security

Monitoring, backup, and security should not be afterthoughts, but necessary processes before deploying MongoDB in a production environment. In addition, monitoring can (and should) be used to troubleshoot and improve performance at the development stage.

In this chapter, we will discuss the operational aspects of MongoDB. Having a backup strategy that produces correct and consistent backups, as well as making sure that our backup strategy will work in the unfortunate case that a backup is needed, will be covered in this chapter. Finally, we will discuss security for MongoDB for many different aspects, such as authentication, authorization, network-level security, and how to audit our security design.

This chapter will focus on the following three areas:

- Monitoring clusters
- Cluster backups
- Securing our clusters

Technical requirements

You will need MongoDB version 5 or a free tier account in MongoDB Atlas installed to smoothly sail through the chapter. The code that has been used for all the chapters and can be found at `https://github.com/PacktPublishing/Mastering-MongoDB-6.x`.

Monitoring clusters

When we are designing a software system, we undertake many explicit and implicit assumptions. We always try to make the best decisions based on our knowledge, but there may be some parameters that we have underestimated or didn't take into account.

Using monitoring, we can validate our assumptions and verify that our application performs as intended and scales as expected. Good monitoring systems are also vital for detecting software bugs and helping us detect early potential security incidents.

The following screenshot shows the wealth of options for monitoring in MongoDB Atlas. Detailing each and every metric is outside the scope of this chapter, but it's important to have at least a cursory understanding of what each metric measures.

Figure 9.1 – MongoDB Atlas monitoring metrics

What should we monitor?

By far the most important metric to monitor in MongoDB is memory usage. MongoDB (and every database system, for what it's worth) uses system memory extensively to increase performance. Whether we use the built-in WiredTiger storage engine or one of the commercial alternatives, memory usage is the first thing that we should keep our eyes on.

Understanding how computer memory works can help us evaluate metrics from our monitoring system. The following are the most important concepts related to computer memory.

Page faults

RAM is fast but expensive. Hard disk drives, or solid-state drives, are relatively cheaper and slower, and they also provide durability for our data in the case of system and power failures. All of our data is stored on the disk, and when we perform a query, MongoDB will try to fetch data from the memory. If the data is not in the memory, then it will fetch the data from the disk and copy it to the memory. This is a **page fault event** because the data in the memory is organized into pages.

As page faults happen, the memory gets filled up and eventually, some pages need to be cleared for more recent data to come into the memory. This is called a **page eviction event**. We cannot completely avoid page faults unless we have a really static dataset, but we do want to try to minimize page faults. This can be achieved by holding our working set in memory.

Memory

The **resident memory** size is the total amount of memory that MongoDB owns in the RAM. This is the base metric to monitor, and it should be less than 80% of the available memory.

Virtual and mapped memory

When MongoDB asks for a memory address, the operating system will return a virtual address. This may or may not be an actual address in the RAM, depending on where the data resides. MongoDB will use this virtual address to request the underlying data. When we have journaling enabled (which should be always enabled, unless we fully understand and accept the data loss risk from disabling it), MongoDB will keep another address on record for the journaled data. Data after the last checkpoint, which is by default every 60 seconds, will be written in the journal. The virtual memory refers to the size of all of the data requested by MongoDB, including the journaling.

The mapped memory excludes journaling references.

What all of this means is that over time, our mapped memory will be roughly equal to our working set, and the virtual memory will be our mapped memory size plus the dataset size after the last checkpoint.

Working sets

The working set is the data size that MongoDB uses. In the case of a transactional database, over time, this will converge to the data size that MongoDB holds, but there may be cases where we have collections that are not used at all and will not contribute to our working set.

We can directly monitor memory in MongoDB Atlas at the MongoDB level using the **Memory** option from *Figure 9.1*. We can indirectly monitor memory usage by using the hardware metrics from the same page named after **System Memory and Max System Memory**.

Tracking free space

A common issue with any database is running out of free disk space. As with the memory, we need to track the disk space usage and be proactive, rather than reactive, with it. Keep monitoring the disk space usage, with proper alerts when it reaches 40%, 60%, or 80% of the disk space, especially for datasets that grow quickly.

Disk space issues are often the ones that cause the most headaches for administrators, DevOps, and developers because of the time it takes to move data around.

> **Note**
>
> The `directoryperdb` option can help with data sizing as we can split our storage into different physically mounted disks.

We can monitor **disk IOPS and disk free space** in the **Hardware Metrics** section of the monitoring dashboard in MongoDB Atlas.

Monitoring replication

Replica sets use the **operations log (oplog)** to keep the synced state. Every operation gets applied on the primary server and then gets written in the primary server's oplog, which is a capped collection. Secondaries read this oplog asynchronously and apply the operations one by one.

If the primary server gets overloaded, then the secondaries won't be able to read and apply the operations fast enough, generating replication lag. **Replication lag** is counted as the time difference between the last operation applied on the primary and the last operation applied on the secondary, as stored in the oplog capped collection.

For example, if the time is 4:30 PM and the secondary just applied an operation that was applied on our primary server at 4:25 PM, this means that the secondary is lagging five minutes behind our primary server.

In our production cluster, the replication lag should be close to (or equal to) zero.

We can monitor the **replication lag** in the **MongoDB Metrics** section of the monitoring dashboard in MongoDB Atlas.

Oplog size

Every member in a replica size will have a copy of the oplog in `db.oplog.rs()`. The reason for this is that if the primary steps down, one of the secondaries will get elected, and it needs to have an up-to-date version of the oplog for the new secondaries to track.

The oplog size is configurable, and we should set it to be as large as possible. The oplog size doesn't affect the memory usage and can make or break the database in cases of operational issues.

The reason for this is that if the replication lag increases over time, we will eventually get to the point where the secondaries will fall so far behind the primary that the secondary server won't be able to read from the primary's oplog; this is because the oldest entry in the primary's oplog will be later than the latest entry that was applied in our secondary server. Essentially, in this case, there will be a gap between the secondary and primary oplogs. During this gap, we lost track of operations that might have happened, and this causes MongoDB to halt replication.

In general, the oplog should hold at least 1 to 2 days' worth of operations. The oplog should be longer than the time it takes for the initial sync, for the same reason that is detailed earlier.

We can monitor the **oplog size** in the MongoDB Metrics section of monitoring in MongoDB Atlas using the **Replication Oplog** window and the **Oplog GB/Hour** metrics.

Network

Network usage can be the bottleneck in performance or it can be indicative of another issue with the environment. In both cases, it's useful to keep an eye on network usage and if possible report any anomalies for further investigation. We can inspect **Bytes in**, **Bytes out**, and **number of requests** using the **Network** option in the **MongoDB Metrics** section of the monitoring dashboard in MongoDB Atlas.

Cursors and connections

The number of **connections** and the number of **cursors** (open and timed out) in the MongoDB Metrics section of monitoring in MongoDB Atlas can verify the health state of our cluster.

Too many connections may drain the database and reveal a systemic issue with our application code. Too few connections can be underutilizing our database.

Too many open cursors will also put extra load on our database, and too many timeout cursors means that our database is already under heavy load.

Document Metrics

The **Document Metrics** option in the **MongoDB Metrics** section of the monitoring dashboard in MongoDB Atlas outputs the number of CRUD operations happening every second across each server in the cluster.

Not all CRUD operations are created equal in terms of adding load to the database, but on average, this is a useful metric to identify if one of the servers is behaving out of sync with the rest of the cluster.

For example, if we have two identical secondaries in a replica set and one of them is showing less than half of the operations occurring compared to the other, it could be a sign of some network bandwidth or a configuration issue between this server and the primary.

Monitoring memory usage in WiredTiger

Using WiredTiger, we can define the internal cache memory usage on startup. By default, the internal cache will be ((total RAM size in GB) - 1) / 2, with a lower limit of 256 MB and an upper limit of 10 GB.

This means that in a system with 16 GB RAM, the internal cache size would be (16 - 1) / 2 = 7.5 GB.

On top of the internal cache, there is also memory that MongoDB can allocate for other operations such as maintaining connections and data processing (in-memory sort, map-reduce, aggregation, and more).

MongoDB processes using the WiredTiger storage engine will also use all of the available memory in the underlying operating system by using the filesystem cache.

We can view the settings for the WiredTiger cache via the mongo shell as follows:

```
> db.serverStatus().wiredTiger.cache
```

We can adjust its size by using the `storage.wiredTiger.engineConfig.cacheSizeGB` parameter.

The generic recommendation is to leave the WiredTiger internal cache size at its default. If our data has a high compression ratio, it may be worth reducing the internal cache size by 10% to 20% to free up more memory for the filesystem cache.

Tracking page faults

The number of page faults can remain fairly stable and not affect performance significantly. However, once the number of page faults reaches a certain threshold, our system will be quickly and severely degraded. This is even more evident for HDDs, but it affects **solid-state drives (SSDs)** as well.

The way to ensure that we don't run into problems regarding page faults is to always have a staging environment that is identical to our production in setup. This environment can be used to stress test how many page faults our system can handle, without deteriorating performance. Comparing the actual number of page faults in our production system with the maximum number of page faults that we calculated from our staging system, we can find out how much leeway we have left.

Another way to view page faults is via the shell, looking at the `extra_info` field of the `serverStatus` output as follows:

```
> db.adminCommand({"serverStatus" : 1})['extra_info']
{ "note" : "fields vary by platform", "page_faults" :
Long("3465") }
```

As the `note` states, these fields may not be present on every platform.

Tracking B-tree misses

As you saw in the previous chapter, proper indexing is the best way to keep MongoDB responsive and performant. B-tree misses refer to page faults that happen when we try to access a B-tree index. Indexes are usually used frequently and are relatively small compared to our working set and the memory available, so they should be in the memory at all times.

If we have an increasing number of B-tree misses or ratio of B-tree hits, or if there is a decrease in the number of B-tree misses, it's a sign that our indexes have grown in size and/or are not optimally designed. B-tree misses can also be monitored via MongoDB Cloud Manager, or in the shell.

> **Note**
> In the shell, we can use collection stats to get B-tree information, located under the B-tree subdocument.

I/O wait

I/O wait refers to the time that the operating system waits for an I/O operation to complete. It has a strong positive correlation with page faults. If we see I/O wait increasing over time, it's a strong indication that page faults will follow as well. We should aim to keep the I/O wait at less than 60% to 70% for a healthy operational cluster. This can buy us some time to upgrade in the case of a suddenly increased load.

Read and write queues

Another way to look at I/O wait and page faults is via read and write queues. When we have page faults and I/O wait, requests will inevitably start to queue for either reads or writes. Queues are the effect rather than the root cause, so by the time the queues start building up, we know we have a problem to solve.

Lock percentage

Lock percentage was important with the original MMAPv1 MongoDB storage engine, which was deprecated in MongoDB version 4. WiredTiger implements document-level concurrency and transparently retries operations to overcome write conflicts.

WiredTiger only needs to use an intent lock in the global, database, and collection levels and so locking is not nearly as much of an issue in the current version as it was in the past.

While locking is less of an issue, we can still have high contention in any of the document, collection, database, or global levels that will result in queues building up over time. We can monitor the number of operations waiting for a read lock, a write lock, or the total (sum) of these two. These metrics and many others are available in every MongoDB environment. For example, in MongoDB Atlas, **queues** is currently an option under the **View monitoring** option on the cluster dashboard page.

Working set calculations

A working set is the strongest indicator of our memory requirements. Ideally, we would like to have our entire dataset in the memory, but most of the time, this is not feasible. The next best thing is to have our working set in memory. The working set can only be calculated indirectly as of MongoDB version 3.0.

Indirectly, our working set is the size of the data that we need to satisfy 95% or more of our user's requests. To calculate this, we need to identify the queries that the users make and which datasets they use from the logs. Adding 30% to 50% to it for index memory requirements, we can arrive at the working set calculation.

Another indirect way of estimating the working size is through the number of page faults. If we don't have page faults, then our working set fits in the memory. Through trial and error, we can estimate the point at which the page faults start to happen and understand how much more of a load our system can handle.

If we can't have the working set in memory, then we should have at least enough memory so that the indexes can fit in memory. In the previous chapter, we described how we can calculate index memory requirements and how we can use this calculation to size our RAM accordingly.

Monitoring tools

There are several options for monitoring. In this section, we will discuss how we can monitor by using MongoDB's own tools or third-party tools.

Free cloud-based monitoring

MongoDB offers free cloud-based monitoring for on-premises installations. It is as simple as typing `db.enableFreeMonitoring()` in the command line.

The result then will be as follows:

```
{
...
"url": "<MONGODB_MONITORING_URL>"
...
}
```

We can then go to the `MONGODB_MONITORING_URL` and watch metrics from our server. This URL is not authenticated and it's only hard to guess because of a long unique part of the URL. This is fair enough for development purposes but should not be used for any production deployment as we could potentially leak our cluster's monitoring status to an adversary.

SaaS-hosted tools

MongoDB, Inc.'s own tool MongoDB Cloud Manager (formerly MongoDB Monitoring Service) is a robust tool for monitoring all of the metrics that were described earlier. MongoDB Cloud Manager requires a MongoDB Enterprise Advanced subscription with a 30-day trial period.

MongoDB Cloud Manager metrics can also be accessed via MongoDB Atlas, MongoDB, Inc.'s DBaaS offering. This also has a limited free tier and is available in all three major cloud providers (Amazon, Google, and Microsoft).

On-premises tools

MongoDB Ops Manager (available with MongoDB Enterprise Advanced), is the on-premises version of the MongoDB Cloud Manager.

MongoDB Cloud Manager should be used when we want a SaaS-hosted tool to monitor our own MongoDB servers, deployed on-premises or in the cloud of our choice.

MongoDB Ops Manager should be used where we can't have a SaaS pointing to our servers. This could be, for example, because of security requirements.

We can use the Ops Manager (as part of downloading MongoDB Enterprise Advanced) in our local dev environment for free for evaluation and development purposes, but any other production use, including testing (pre-prod environments), is forbidden by the Terms of Use of MongoDB Enterprise Advanced.

Open source tools

All major open source tools, such as **Nagios**, **Munin**, and **Cacti**, provide plugin support for MongoDB. Although it is beyond the scope of this book, operations and DevOps should be familiar with both setting up and understanding the metrics that were described earlier in order to effectively troubleshoot MongoDB and preemptively resolve issues before they grow out of proportion.

The `mongotop` and `mongostat` commands and scripts in the mongo shell can also be used for ad hoc monitoring. One of the risks with such manual processes, however, is that any failure of the scripts may jeopardize our database. If there are well-known and tested tools for your monitoring needs, please avoid writing your own.

In this section, we went through the attributes that we should be monitoring using MongoDB. We also learned about an array of different tools that we can use to monitor our MongoDB database, both on-premises or in the cloud. In the following section, we will learn about one of the best safeguards for when (and not if) things go south: taking backups of our data.

Cluster backups

A well-known maxim goes as follows:

"Hope for the best, plan for the worst."

– John Jay (1813)

This should be our approach when designing our backup strategy for MongoDB. There are several distinct failure events that can happen.

Backups should be the cornerstone of our disaster recovery strategy in case something happens. Some developers may rely on replication for disaster recovery, as it seems that having three copies of our data is more than enough. We can always rebuild the cluster from the other two copies in case one of the copies is lost.

This is the case in the event of disks failing. Disk failure is one of the most common failures in a production cluster and will statistically happen once the disks start reaching their **mean time between failures (MTBF)**.

However, it is not the only failure event that can happen. Security incidents, or human errors, are just as likely to happen and should be an integral part of our plan. Catastrophic failures by means of losing all replica set members at once, from a fire, a flood, an earthquake, or a disgruntled employee, are events that should not lead to production data loss.

> **Note**
>
> A useful interim option, in the middle ground between replication and implementing proper backups, could be setting up a delayed replica set member. This member can lag several hours or days behind the primary server so that it will not be affected by malicious changes in the primary. The important detail to take into account is that the oplog needs to be configured so that it can hold several hours of delay. Also, this solution is only an interim as it doesn't take into account the full range of reasons why we need disaster recovery, but it can definitely help with a subset of them.

Disaster recovery is a class of failures that require backups to be taken not only periodically, but also by using a process that isolates them (both geographically and in terms of access rules) from our production data.

Backup options

Depending on our deployment strategy, we can choose different options for backups.

Cloud-based solutions

The most straightforward solution arises if we are using a cloud DBaaS solution. In the example of MongoDB Atlas, we can manage backups from the GUI or using the API.

MongoDB Atlas offers the Cloud Backups service, which uses the snapshot functionality from the underlying cloud provider (AWS, Microsoft Azure, or Google Cloud Platform) to provide both **on-demand** and **Continuous Cloud Backups** with a frequency and retention that is dependent on the MongoDB Atlas cloud level of service selected. This feature is only available on paid plans.

Continuous Cloud Backups use the oplog to back up our data, so we need to make sure that the oplog size is larger than the writes that will incur between two backup attempts. If the oplog size is not large enough, then the backup will be aborted.

Backups, either on-demand or continuously taken, guarantee point-in-time recovery.

> **Note**
>
> MongoDB Atlas backups are the recommended method by MongoDB for sharded clusters, only because of their operational efficiency.

On-premises solutions

If we host MongoDB on our own servers, we can use MongoDB, Inc.'s MongoDB Cloud Manager. Cloud Manager is a SaaS that we can point to our own servers to monitor and back up our data. It uses the same oplog that replication uses and can back up both replica sets and sharded clusters.

If we don't want to (or can't, for security reasons) point our servers to an external SaaS service, we can use MongoDB Cloud Manager's functionality on-premises, using MongoDB Ops Manager. To get MongoDB Ops Manager, we need to get a subscription to the Enterprise Advanced edition of MongoDB for our cluster.

Backups with filesystem snapshots

The most common backup method in the past, and one that is still sometimes used, relies on the underlying filesystem point-in-time snapshots functionality to back up our data.

EBS on EC2, and **Logical Volume Manager (LVM)** on Linux, support point-in-time snapshots.

> **Note**
>
> If we use WiredTiger with the latest version of MongoDB, we can have volume-level backups, even if our data and journal files reside in different volumes.

We can make a backup of a replica set as follows:

- To make a backup of a replica set, we need to have a consistent state for our database. This implies that we have all of our writes either committed to the disk or in our journal files.

- If we use WiredTiger storage, our snapshot will be consistent as of the latest checkpoint, which is either 2 GB of data or the last-minute backup.

> **Note**
>
> Ensure that you store the snapshot in an off-site volume for disaster recovery purposes. You need to have enabled journaling to use point-in-time snapshots. It's good practice to enable journaling regardless.

Making a backup of a sharded cluster

If we want to make a backup of an entire sharded cluster, we need to stop the balancer before starting. The reason is that if there are chunks migrating between different shards at the time that we take our snapshot, our database will be in an inconsistent state, having either incomplete or duplicate data chunks that were in flight at the time we took our snapshot.

Backups from an entire sharded cluster will be approximate in time. If we need point-in-time precision, we need to stop all of the writes in our database, something that is generally not possible for production systems.

First, we need to disable the balancer by connecting to mongos through the mongo shell as follows:

```
> use config
> sh.stopBalancer()
```

Then, if we don't have journaling enabled in our secondaries, or if we have journal and data files in different volumes, we need to lock our secondary mongo instances for all shards and the config server replica set.

> **Note**
>
> We also need to have a sufficient oplog size in these servers so that they can catch up to the primaries once we unlock them; otherwise, we will need to resync them from scratch.

Given that we don't need to lock our secondaries, the next step is to back up the config server. In Linux (and using LVM), this would be similar to doing the following:

```
$ lvcreate --size 100M --snapshot --name snap-14082017 /dev/
vg0/mongodb
```

Then, we need to repeat the same process for a single member from each replica set in each shard.

Finally, we need to restart the balancer using the same mongo shell that we used to stop it, as follows:

```
> sh.setBalancerState(true)
```

Without going into too much detail here, it's evident that making a backup of a sharded cluster is a complicated and time-consuming procedure. It needs prior planning and extensive testing to make sure that it not only works with minimal disruption but also that our backups are usable and can be restored back to our cluster.

> **Note**
>
> MongoDB Atlas can back up an entire sharded cluster without any of the preceding steps. It's all possible using the GUI or API, providing a guarantee for point-in-time restore of a snapshot. It is the recommended method by MongoDB.

Making backups using mongodump

The mongodump tool is a command-line tool that can make a backup of the data in our MongoDB cluster. As such, the downside is that all of the indexes need to be recreated on restore, which may be a time-consuming operation.

The major downside that the mongodump tool has is that in order to write data to the disk, it needs to bring data from the internal MongoDB storage to the memory first. This means that in the case of production clusters running under strain, mongodump will invalidate the data residing in the memory from the working set with the data that would not be residing in the memory under regular operations. This degrades the performance of our cluster.

On the plus side, when we use mongodump, we can continue taking writes in our cluster, and if we have a replica set, we can use the --oplog option to include the entries that occur during the mongodump operation in its output oplog.

If we go with that option, we need to use --oplogReplay when we use the mongorestore tool to restore our data back to the MongoDB cluster.

mongodump is a great tool for single-server deployments, but once we get to larger deployments, we should consider using different (and better planned) approaches to back up our data.

> **Note**
>
> We can't use mongodump/mongorestore with a sharded cluster in MongoDB 4.2 or later. This is because in-flight sharded transactions may end up in an inconsistent state.

Backing up by copying raw files

If we don't want to use any of the preceding options that were outlined, our last resort is to copy the raw files using cp/rsync, or something equivalent. This is generally not recommended for the following reasons:

- We need to stop all of the writes before copying files.
- The backup size will be larger since we need to copy indexes and any underlying padding and fragmentation storage overhead.
- We cannot get point-in-time recovery by using this method for replica sets, and copying data from sharded clusters in a consistent and predictable manner is extremely difficult.

> **Note**
>
> Making a backup by copying raw files should be avoided unless no other option really exists.

Making backups using queuing

Another strategy that's used in practice is utilizing a queuing system, intercepting our database and the frontend software system. Having something such as an ActiveMQ queue before the inserts/updates/deletes in our database means that we can safely send out data to different sinks, which are MongoDB servers or log files in a separate repository. Like the delayed replica set method, this method can be useful for a class of backup problems but can fail for some others.

> **Note**
>
> This is a useful interim solution, but it should not be used as a permanent one.

EC2 backup and restore

MongoDB Cloud Manager can automate making backups from EC2 volumes. Given that our data is in the cloud, we could use the Cloud Manager instead.

But then, if we can't use the Cloud Manager for some reason, we can write a script to make a backup by implementing the following steps:

1. Assuming that we have journaling enabled (and we really should) and we have already mapped `dbpath`, containing data and journal files to a single EBS volume, we first need to find the EBS block instances associated with the running instance by using `ec2-describe-instances`.

2. The next step is to find the logical volumes that `dbpath` of our MongoDB database is mapped to using `lvdisplay`.

3. Once we have identified the logical devices from the logical volumes, we can use `ec2-create-snapshot` to create new snapshots. We need to include each and every logical device that maps to our `dbpath` directory.

To verify that our backups work, we need to create new volumes based on the snapshots and mount the new volumes there. Finally, the `mongod` process should be able to start mounting the new data, and we should connect by using MongoDB to verify these.

Using EC2 servers to host our MongoDB servers is less popular now than it was in the past, mostly because of the wide variety of other options available.

Incremental backups

Making full backups every time may be viable for some deployments, but as the size reaches a certain threshold, full backups take too much time and space.

At this point, we will want to make full backups every once in a while (maybe once per month, for example) and incremental backups in-between (for example, nightly).

Both Ops Manager and Cloud Manager support incremental backups, and if we get to this size, it may be a good idea to use a tool to make our backups instead of rolling out our own.

If we don't want to (or can't) use these tools, we have the option of restoring via the oplog as follows:

1. Make a full backup with any method that was described earlier.

2. Lock writes on the secondary server of our replica set.

3. Note the latest entry in the oplog.

4. Export the entries from the oplog after the latest entry as follows:

    ```
    > mongodump --host <secondary> -d local -c oplog.rs -o /
    mnt/mongo-oldway_backup
      --query '{ "ts" : { $gt : Timestamp(1467999203, 391) }
    }'
    ```

5. Unlock writes on the secondary server.

To restore, we can use the `oplog.rs` file that we just exported, and use `mongorestore` with the `--oplogReplay` option as follows:

```
> mongorestore -h <primary> --port <port> --oplogReplay <data_
file_position>
```

This method requires locking writes and as such is not recommended in general.

An even better solution is to use the LVM filesystem with incremental backups, but this depends on the underlying LVM implementation, which we may or may not be able to tweak.

Making backups is the best way to protect ourselves from catastrophic failures, no matter whether the root cause is in our control or not. In the following section, we will learn about how we can best secure our data from unauthorized access. Security is one of the most important non-functional requirements that we must take into account while designing, developing, and maintaining MongoDB-based systems.

Securing our clusters

Security is a multifaceted goal in a MongoDB cluster. For the rest of this chapter, we will examine different attack vectors and how we can protect against them. In addition to these best practices, developers and administrators must always use common sense so that security interferes only as much as is required for operational goals.

Authentication

Authentication refers to verifying the identity of a client. This prevents the impersonation of someone in order to gain access to their data.

Username/password client-based authentication

The simplest way to authenticate is by using a `username` and `password` pair. This can be done via the shell in two ways, the first of which is as follows:

```
> db.auth( <username>, <password> )
```

Passing in a comma-separated `username` and `password` will assume the default values for the rest of the fields as follows:

```
> db.auth( {
  user: <username>,
  pwd: <password>,
  mechanism: <authentication mechanism>,
  digestPassword: <boolean>
} )
```

If we pass a document object, we can define more parameters than `username/password`.

Client-server and server-server key-based authentication

Authenticating our clients with a `username/password` pair is the simplest way to authenticate our clients to the server.

Taking this one step forward, we can authenticate our users with a certificate file. The (authentication) `mechanism` parameter can take several different values, with the default being `SCRAM-SHA-256`.

MONGODB-x.509 is used for TLS/SSL authentication. Users and internal replica set servers can be authenticated by using SSL certificates, which are self-generated and signed or come from a trusted third-party authority.

To configure x.509 for internal authentication of replica set members, we need to supply one of the following parameters.

The following is for the configuration file:

```
security.clusterAuthMode / net.ssl.clusterFile
```

The following is used on the command line:

```
--clusterAuthMode and --sslClusterFile
> mongod --replSet <name> --sslMode requireSSL
--clusterAuthMode x509 --sslClusterFile <path to membership
certificate and key PEM file> --sslPEMKeyFile <path to SSL
certificate and key PEM file> --sslCAFile <path to root CA PEM
file>
```

MongoDB Enterprise Advanced, the paid offering from MongoDB, Inc., adds two more options for authentication as follows:

- The first added option is the **Generic Security Service Application Program Interface (GSSAPI)** Kerberos. Kerberos is a mature and robust authentication system that can be used for Windows-based Active Directory deployments, among others.

- The second added option is **LDAP Proxy Authentication (LDAP SASL)**. LDAP is similar to Kerberos; it's a mature and robust authentication mechanism that is the foundation behind Microsoft's Active Directory.

We can mutually authenticate our replica set or sharded cluster servers to one another using the same methodology, and this is good practice to do in a production environment.

Localhost exception

Landing into a project that's already well on the way should have authentication and roles sorted already. It's only when we are starting a greenfield project that we need to use the localhost exception. Simply put, the localhost exception allows us to create only the first user or role ever in the database.

In a standalone database or a replica set, the localhost exception rule only applies when there are no users or roles set up.

In a sharded environment, the localhost exception rule applies to each shard in isolation and the entire cluster as a whole.

> **Note**
>
> We should create the first user using either the `userAdmin` or `userAdminAnyDatabase` role to be able to grant any permissions and create new users through this root user.

MongoDB Atlas also supports passwordless authentication using AWS IAM roles. It is compatible with passwordless AWS IAM authenticatiion setup or we can use SAML Single Sign On (SSO) using any 3rd party provider such as for example, Okta or Ping. The secret key that we will use locally for authentication is not persisted by the local driver or transmitted over the wire to the MongoDB Atlas servers, which is convenient for security reasons.

Role-based access control-based authorization

After we have configured the authentication to verify that the users are who they claim they are when connecting to our MongoDB server, we need to configure the rights that each one of them will have in our database.

This is the **authorization** aspect of permissions. MongoDB uses role-based access control to control permissions for different user classes.

Every role has permission to perform some actions on a resource.

A resource can be one or more collections or databases.

The command's format is as follows:

```
{ db: <database>, collection: <collection> }
```

If we specify " " (an empty string) for either db or collection, it means any db or collection. An example of this is as follows:

```
{ db: "mongo_books", collection: "" }
```

This would apply our action in every collection in the mongo_books database.

> **Note**
>
> If the database is not the admin database, this will not include the system collections. System collections, such as <db>.system.profile, <db>.system.js, admin.system. users, and admin.system.roles, need to be defined explicitly.

Similar to the preceding option, we can define the following:

```
{ db: "", collection: "" }
```

We define this to apply our rule to all of the collections across all of the databases, except for system collections, of course.

We can also apply rules across an entire cluster as follows:

```
{ resource: { cluster : true }, actions: [ "addShard" ] }
```

The preceding example grants privileges for the addShard action (adding a new shard to our system) across the entire cluster. The cluster resource can only be used for actions that affect the entire cluster, rather than a collection or database (for example, shutdown, replSetReconfig, appendOplogNote, resync, closeAllDatabases, and addShard).

What follows is an extensive list of cluster-specific actions and some of the most widely used actions.

The list of the most widely used actions is as follows:

- find
- insert
- remove
- update
- bypassDocumentValidation
- viewRole/viewUser
- createRole/dropRole
- createUser/dropUser
- inprog
- killop
- replSetGetConfig/replSetConfigure/replSetStateChange/resync
- getShardMap/getShardVersion/listShards/moveChunk/removeShard/addShard
- dropDatabase/dropIndex/fsync/repairDatabase/shutDown
- serverStatus/top/validate

The cluster-specific actions are as follows:

- unlock
- authSchemaUpgrade
- cleanupOrphaned
- cpuProfiler
- inprog
- invalidateUserCache
- killop
- appendOplogNote

- `replSetConfigure`
- `replSetGetConfig`
- `replSetGetStatus`
- `replSetHeartbeat`
- `replSetStateChange`
- `resync`
- `addShard`
- `flushRouterConfig`
- `getShardMap`
- `listShards`
- `removeShard`
- `shardingState`
- `applicationMessage`
- `closeAllDatabases`
- `connPoolSync`
- `fsync`
- `getParameter`
- `hostInfo`
- `logRotate`
- `setParameter`
- `shutdown`
- `touch`
- `connPoolStats`
- `cursorInfo`
- `diagLogging`
- `getCmdLineOpts`
- `getLog`
- `listDatabases`
- `netstat`

- `serverStatus`

- `top`

If this sounds too complicated, that's because it is! The flexibility that MongoDB allows for configuring different actions on resources means that we need to study and understand the extensive lists, as described earlier.

Thankfully, some of the most common actions and resources are bundled in built-in roles.

We can use these built-in roles to establish the baseline of permissions that we will give to our users, and then fine-grain these based on the extensive list.

User roles

There are two different generic user roles that we can specify as follows:

- `read`: A read-only role across non-system collections and the following system collections: `system.indexes`, `system.js`, and `system.namespaces` collections

- `readWrite`: A read and modify role across non-system collections and the `system.js` collection

Database administration roles

There are three database-specific administration roles as follows:

- `dbAdmin`: The basic admin user role that can perform schema-related tasks, indexing, and gathering statistics. A `dbAdmin` cannot perform user and role management.

- `userAdmin`: Create and modify roles and users. This is complementary to the `dbAdmin` role.

> **Note**
>
> A `userAdmin` can modify itself to become a superuser in the database, or, if scoped to the `admin` database, the MongoDB cluster.

- `dbOwner`: Combining `readWrite`, `dbAdmin`, and `userAdmin` roles, this is the most powerful admin user role.

Cluster administration roles

The following are the cluster-wide administration roles that are available:

- `hostManager`: Monitor and manage servers in a cluster.

- `clusterManager`: Provides management and monitoring actions on the cluster. A user with this role can access the config and local databases, which are used in sharding and replication, respectively.

- `clusterMonitor`: Read-only access for monitoring tools provided by MongoDB, such as MongoDB Cloud Manager and the Ops Manager agent.

- `clusterAdmin`: Provides the greatest cluster management access. This role combines the privileges that are granted by the `clusterManager`, `clusterMonitor`, and `hostManager` roles. Additionally, the role provides the `dropDatabase` action.

Backup and restore roles

Role-based authorization roles can be defined in the backup and restore granularity level as follows:

- `backup`: Provides privileges that are needed to back up the data. This role provides sufficient privileges to use the MongoDB Cloud Manager backup agent, the Ops Manager backup agent, or `mongodump`.

- `restore`: Provides the privileges that are needed to restore data with `mongorestore`, without the `--oplogReplay` option or `system.profile` collection data.

Roles across all databases

Similarly, the following is the set of available roles across all databases:

- `readAnyDatabase`: Provides the same read-only permissions as `read`, except it applies to all but the local and config databases in the cluster. The role also provides the `listDatabases` action on the cluster as a whole.

- `readWriteAnyDatabase`: Provides the same read and write permissions as `readWrite`, except it applies to all but the local and config databases in the cluster. The role also provides the `listDatabases` action on the cluster as a whole.

- `userAdminAnyDatabase`: Provides the same access to user administration operations as `userAdmin`, except it applies to all but the local and config databases in the cluster. Since the `userAdminAnyDatabase` role allows users to grant any privilege to any user, including themselves, the role also indirectly provides superuser access.

- `dbAdminAnyDatabase`: Provides the same access to database administration operations as `dbAdmin`, except it applies to all but the local and config databases in the cluster. The role also provides the `listDatabases` action on the cluster as a whole.

Superuser roles

Finally, the following are the available superuser roles:

- `root`: Provides access to the operations and all of the resources of the `readWriteAnyDatabase`, `dbAdminAnyDatabase`, `userAdminAnyDatabase`, `clusterAdmin`, `restore`, and `backup` combined.

- __system: The __system role is assigned by MongoDB internally and is not designed to be used by applications or users. This role grants database-wide unrestricted access. If you definitely need this role, it's better to create a user-defined role as explained earlier and grant it anyAction on anyResource instead.

> **Note**
> Superuser roles should be avoided as they can have potentially destructive permissions across all of the databases on our server.

Network-level security

Apart from MongoDB-specific security measures, the following are the best practices that have been established for network-level security:

- Only allow communication between servers, and only open the ports that are used for communicating between them. Use a firewall to ensure that every other port is closed.

- Always use TLS/SSL for communication between servers. This prevents man-in-the-middle attacks from impersonating a client.

- Always use different sets of development, staging, and production environments and security credentials. Ideally, create different accounts for each environment, and enable two-factor authentication in both staging and production environments.

- Use a VPN over TLS/SSL (for best performance) if we need to communicate to a server in a remote location. An example of such usage could be a delayed secondary server in a replica set in a remote location used for disaster recovery purposes.

- Use the net.bindIp configuration to limit communication to the mongod or mongos processes to the backend servers that will process the front end (UI) requests.

Auditing security

No matter how much we plan our security measures, a second or third pair of eyes from someone outside of our organization can give a different view of our security measures and uncover problems that we may have underestimated or overlooked. Don't hesitate to involve security experts and white hat hackers to do penetration testing on your servers.

Special cases

Medical or financial applications require added levels of security for data privacy reasons.

If we are building an application in the healthcare space, accessing users' sensitive and personally identifiable information, we may need to get HIPAA certified.

If we are building an application that interacts with payments and manages cardholder information, we may need to become PCI/DSS compliant.

> **Note**
>
> The specifics of each certification are outside the scope of this book, but it is important to know that MongoDB has use cases in these fields that fulfill the requirements, and, as such, it can be the right tool with proper, well-thought design.

Overview

Summing up the best practice recommendations involving security, we have the following:

- **Enforce authentication**: Always enable authentication in production environments.

- **Enable access control**: First, create a system administrator, and then use that administrator to create more limited users. Give as few permissions as needed for each user role.

- **Define fine-grained roles in access control**: Do not give more permissions than needed for each user.

- **Encrypt communication between clients and servers**: Always use TLS/SSL for communication between clients and servers in production environments. Always use TLS/SSL for communication between mongod and mongos or config servers as well.

- **Encrypt data at rest**: The MongoDB Enterprise Advanced edition offers the functionality to encrypt data when stored, using WiredTiger encryption at rest.

> **Note**
>
> Alternatively, we can encrypt data using filesystem, device, or physical encryption. In the cloud, we often get the option for encryption as well (for example, with EBS on Amazon EC2).

- **Limit network exposure**: MongoDB servers should only be connected to the application servers and any other servers that are needed for operations. Ports other than the ones that we set up for MongoDB communications should not be open to the outside world. If we want to debug MongoDB usage, it's important to have a proxy server with controlled access set up to communicate with our database.

- **Audit servers for unusual activity**: The MongoDB Enterprise Advanced edition offers a utility for auditing. By using it, we can output events to the console, a JSON file, a BSON file, or the syslog. In any case, it's important to make sure that audit events are stored in a partition that is not available to the system's users.

- Use a dedicated operating system user to run MongoDB. Make sure that the dedicated operating system user can access MongoDB but doesn't have unnecessary permissions.

- Disable JavaScript server-side scripts if they are not needed.

MongoDB can use JavaScript for server-side scripts with the following commands: `mapReduce()`, `group()`, `$accumulator`, and `$function`. If we don't need these commands, we should disable server-side scripting by using the `–noscripting` option on the command line or setting `security.javascriptEnabled=false`.

Summary

In this chapter, you learned about three operational aspects of MongoDB: monitoring, backup, and security.

We discussed the metrics that we should monitor in MongoDB and how to monitor them. Following that, we discussed how to make backups and ensure that we can use them to restore our data. Finally, you learned about security with authentication and authorization concepts as well as network-level security and how to audit it.

As important as it is to design, build, and extend our application as needed, it is equally important to make sure that we have peace of mind during operations and are safeguarded from unexpected events, such as human error and internal or external malicious users.

In the following chapter, you will learn about pluggable storage engines, a new concept that was introduced in version 3.0 of MongoDB. Pluggable storage engines allow different use cases to be served, especially in application domains that have specific and stringent requirements concerning data handling and privacy.

10
Managing Storage Engines

MongoDB introduced the concept of pluggable storage engines in version 3.0. After the acquisition of WiredTiger, it introduced its storage engine as optional at first, and then as the default storage engine for the current version of MongoDB. In this chapter, we will deep dive into the concept of storage engines, why they matter, and how we can choose the best one according to our workload.

We will also discuss the concept of database locking, why it used to be important, and how it is becoming increasingly less important with the WiredTiger storage engine.

In this chapter, we will cover the following topics:

- Pluggable storage engines
- Locking in MongoDB

Pluggable storage engines

With MongoDB breaking out from the web application paradigm into domains with different requirements, storage has become an increasingly important consideration.

Using multiple storage engines can be seen as an alternative way to using different storage solutions and databases in our infrastructure stack. This way, we can reduce operational complexity and development time to market since the application layer is agnostic of the underlying storage layer.

At the time of writing, MongoDB offers four different storage engines. We will look at these in more detail in the following sections.

WiredTiger

As of version 3.2, WiredTiger is the default storage engine, and also the best choice for most workloads. By providing document-level locking, it overcomes one of the most significant drawbacks earlier versions of MongoDB had – lock contention under high load.

We will explore some of WiredTiger's benefits in the following sections.

Document-level locking

Locking is so important that we will explain the performance implications that fine-grained locking has in further detail at the end of this section. Having document-level locking as opposed to MMAPv1 collection-level locking can make a huge difference in many real-world use cases. It is one of the main reasons to choose WiredTiger over MMAPv1.

Snapshots and checkpoints

WiredTiger uses **Multi-Version Concurrency Control (MVCC)**. MVCC is based on the concept that the database keeps multiple versions of an object so that readers will be able to view consistent data that doesn't change during a read.

In a database, if we have multiple readers accessing data at the same time where writers are modifying the data, we can end up with a case where readers view an inconsistent view of this data. The simplest and easiest way to solve this problem is to block all readers until the writers are done modifying data.

This will, of course, cause severe performance degradation. MVCC solves this problem by providing a snapshot of the database for each reader. When the read starts, each reader is guaranteed to view the data exactly as it was when the read started. Any changes made by writers will only be seen by readers after the write has been completed, or, in database terms, after the transaction is committed.

To achieve this goal, when a write is coming in, updated data will be kept in a separate location on disk and MongoDB will mark the affected document as obsolete. MVCC is said to provide point-in-time consistent views. This is equivalent to a read-committed isolation level in traditional RDBMS systems.

For every operation, WiredTiger will snapshot our data at the exact moment that the operation happens and provide a consistent view of application data to the application. When we write data, WiredTiger will create a snapshot for every 2 GB of journal data or every 60 seconds, whichever comes first. WiredTiger relies on its built-in journal to recover any data after the latest checkpoint in case of failure.

> **Note**
> We can disable journaling using WiredTiger, but if the server crashes, we will lose any data after the last checkpoint is written.

Journaling

As explained in the *Snapshots and checkpoints* section, journaling is the cornerstone of WiredTiger crash recovery protection.

WiredTiger compresses the journal using the snappy compression algorithm. We can use the following code to set a different compression algorithm:

```
storage.wiredTiger.engineConfig.journalCompressor
```

We can also disable journaling for WiredTiger by setting the following to `false`:

```
storage.journal.enabled
```

> **Note**
>
> If we use a replica set, we may be able to recover our data from a secondary that will get elected as a primary and start taking writes if our primary fails. It is recommended to always use journaling unless we understand and can take the risk of suffering through the consequences of not using it.

Data compression

MongoDB uses the snappy compression algorithm by default to compress data and prefixes for indexes. Index-prefixed compression means that identical index key prefixes are stored only once per page of memory. Compression not only reduces our storage footprint but increases I/O operations per second since less data needs to be stored and moved to and from disk. Using more aggressive compression can lead to performance gains if our workload is I/O bound and not CPU bound.

We can define `.zlib` compression instead of snappy or no compression by setting the following parameter to `false`:

```
storage.wiredTiger.collectionConfig.blockCompressor
```

> **Note**
>
> Data compression uses less storage at the expense of CPU. `.zlib` compression achieves better compression at the expense of higher CPU usage, as opposed to the default snappy compression algorithm.

We can disable index prefixes compression by setting the following parameter to `false`:

```
storage.wiredTiger.indexConfig.prefixCompression
```

We can also configure storage per index during creation using the following parameter:

```
{ <storage-engine-name>: <options> }
```

Memory usage

WiredTiger is significantly different from the deprecated MMAPv1 storage system in terms of how it uses RAM. MMAPv1 essentially uses the underlying operating system's filesystem cache to page data from disk to memory and vice versa.

WiredTiger, on the other hand, introduces the new concept of the WiredTiger internal cache.

The WiredTiger internal cache is, by default, the larger of the following:

- 50% of RAM minus 1 GB
- 256 MB

This means that if our server has 8 GB of RAM, we will get the following:

max(7 GB , 256 MB) = WiredTiger will use 3.5 GB of RAM

If our server has 1,512 MB of RAM, we will get the following:

max(256 MB, 256 MB) = WiredTiger will use 256 MB of RAM

Essentially, for any server that has less than 1,512 MB of RAM, WiredTiger will use 256 MB for its internal cache.

We can change the size of the WiredTiger internal cache by setting the following parameter:

```
storage.wiredTiger.engineConfig.cacheSizeGB
```

We can also do this from the command line, like so:

```
--wiredTigerCacheSizeGB
```

Apart from the WiredTiger internal cache, which is uncompressed for higher performance, MongoDB uses the filesystem cache, which is compressed, just like MMAPv1, and will end up using all available memory in most cases.

The WiredTiger internal cache can provide similar performance to in-memory storage. As such, it is important to grow it as much as possible.

We can achieve better performance when using WiredTiger with multi-core processors. This is also a big win compared to MMAPv1, which does not scale as well.

> **Note**
>
> We can, and should, use Docker or other containerization technologies to isolate the mongod processes from others and make sure that we know how much memory each process can, and should, use in a production environment. It is not recommended to increase the WiredTiger internal cache above its default value. The filesystem cache should not be less than 20% of the total RAM.

readConcern

WiredTiger supports multiple readConcern levels. Just like writeConcern, which is supported by every storage engine in MongoDB, with readConcern, we can customize how many servers in a replica set must acknowledge the query results for the document to be returned in the result set.

The most notable options for `readConcern` are as follows:

- `local`: This is the default option. It will return the most recent data from the server. Data may, or may not, have propagated to the other servers in a replica set, and we run the risk of a rollback.

- `linearizable`:

  - Only applicable for reads from the primary server

  - Only applicable in queries that return a single result

  - Data returns satisfy two conditions – `majority` and `writeConcern`

  - Data was acknowledged before the start of the read operation

In addition, if we have set `writeConcernMajorityJournalDefault` to `true`, we are guaranteed that the data won't get rolled back.

If we have set `writeConcernMajorityJournalDefault` to `false`, MongoDB will not wait for the `majority` writes to be durable before acknowledging the write. In this case, our data may be rolled back in the event of losing a member from the replica set. The data that's been returned has already been propagated and acknowledged by the majority of the servers before the read has started.

> **Note**
>
> We need to use `maxTimeMS` when using `linearizable` and `majority` read concern levels in case we can't establish `majority` `writeConcern` to avoid blocking, which means we wait for the response forever. In this case, the operation will return a timeout error.

MMAPv1 is the older storage engine and is considered, in many aspects, as being deprecated, but many deployments may still use it.

> **Note**
>
> `local` and `linearizable` read concerns are available for MMAPv1 as well.

WiredTiger collection-level options

When we create a new collection, we can pass in options to WiredTiger, like this:

```
> db.createCollection(
  "mongo_books",
  { storageEngine: { wiredTiger: { configString: "<key>=<value>"
} } }
)
```

This helps create our `mongo_books` collection with a key-value pair from the available ones that WiredTiger exposes through its API. Some of the most widely used key-value pairs are shown in the following table:

Key	Value
`block_allocation`	Best or first
`allocation_size`	512 bytes through to 128 MB; the default is 4 KB
`block_compressor`	None, `.lz4`, `.snappy`, `.zlib`, `.zstd`, or a custom compressor identifier string, depending on the configuration
`memory_page_max`	512 bytes through to 10 TB; the default 5 MB
`os_cache_max`	An integer greater than or equal to zero; the default is zero

Table 10.1 – WiredTiger key-value pairs

The following has been taken from the definition in the WiredTiger documentation, which is located at `http://source.wiredtiger.com/mongodb-6.0/struct_w_t___s_e_s_s_i_o_n.html`:

```
int WT_SESSION::create()
```

Collection-level options allow for flexibility in configuring storage but should be used with extreme care and after careful testing in development/staging environments.

> **Note**
>
> Collection-level options will be propagated to secondaries if they're applied to a primary in a replica set. `block_compressor` can also be configured from the command line globally for the database using the `--wiredTigerCollectionBlockCompressor` option.

WiredTiger performance strategies

As discussed earlier in this chapter, WiredTiger uses an internal cache to optimize performance. On top of that, there is always the filesystem cache, which the operating system uses to fetch data from disk.

By default, we have 50% of RAM dedicated to the filesystem cache; the remaining 50% is dedicated to WiredTiger's internal cache.

The filesystem cache will keep data compressed as it is stored on disk. The internal cache will decompress it as follows:

- **Strategy 1**: Allocate 80% or more to the internal cache. Our goal here is to fit our working set in WiredTiger's internal cache.

- **Strategy 2**: Allocate 80% or more to the filesystem cache. Our goal here is to avoid using the internal cache as much as possible and rely on the filesystem cache for our needs.

- **Strategy 3**: Use an SSD as the underlying storage for fast seek time and keep defaults at 50-50% allocation.

- **Strategy 4**: Enable compression in our storage layer through MongoDB's configuration to save on storage, and potentially improve our performance by having a smaller working set size.

Our workload will dictate whether we need to deviate from the default (**Strategy 1**) to any of the rest. In general, we should use SSDs wherever possible, and with MongoDB's configurable storage, we can even use SSDs for some of the nodes where we need the best performance and keep HDDs for analytics workloads.

WiredTiger B-tree versus LSM indexes

B-tree is the most common data structure for indexes across different database systems. WiredTiger offers the option to use a **log-structured merge (LSM)** tree instead of a B-tree for indexing.

An LSM tree can provide better performance when we have a workload of random inserts that would otherwise overflow our page cache and start paging in data from disk to keep our index up to date.

LSM indexes can be selected from the command line, like so:

```
$ mongod --wiredTigerIndexConfigString "type=lsm,block_
compressor=zlib"
```

The preceding command chooses lsm as type. Here, block_compressor is zlib for indexes in this mongod instance.

Data migration to WiredTiger

As of MongoDB 5.x, there is no straightforward migration path from MMAPv1 to WiredTiger. We need to mongodump or otherwise export our data from MongoDB and then import it into the newly created standalone server, replica set, or sharded cluster using WiredTiger storage.

In the case of a standalone server, we stop and start the server with WTDBPath, as shown here:

```
$ mongod --storageEngine wiredTiger --dbpath <WTDBPath>
```

In the case of a replica set, we need to shut down and restart our secondary servers one by one. After that, we must restart our primary server using the same parameters as with the standalone server.

In the case of a sharded cluster, we also have to follow the same procedure that we did for a replica set for each of the shards in our cluster.

Encrypted at rest

The encrypted storage engine was added to support a series of special use cases, mostly revolving around finance, retail, healthcare, education, and the government.

We need to have encryption for the rest of our data if we must comply with a set of regulations, including the following:

- PCI DSS for handling credit card information
- HIPAA for healthcare applications
- NIST for the government
- FISMA for the government
- STIG for the government

This can be done in several ways, and cloud service providers, such as EC2, provide EBS storage volumes with built-in encryption. Encrypted storage supports Intel's AES-NI-equipped CPUs to accelerate the encryption/decryption process.

The following encryption algorithms are as follows:

- AES-256, CBC (default)
- AES-256, GCM
- FIPS, FIPS-140-2

Encryption is supported at the page level for better performance. When a change is made in a document, instead of re-encrypting/decrypting the entire underlying file, only the page that is affected gets modified.

Encryption key management is a huge aspect of encrypted storage security. Most of the specifications mentioned previously require key rotation at least once per year.

MongoDB's encrypted storage uses an internal database key per node. This key is wrapped by an external (primary) key that must be used to start the node's mongod process. By using the underlying operating system's protection mechanisms, such as mlock or VirtualLock, MongoDB can guarantee that the external key will never be leaked from memory to disk by page faults.

The external (primary) key can be managed either by using the **Key Management Interoperability Protocol (KMIP)** or by using local key management via a key file.

MongoDB can achieve key rotation by performing rolling restarts of the replica set members. Using KMIP, MongoDB can only rotate the external key and not the underlying database files. This delivers significant performance benefits.

> **Note**
>
> Using KMIP is the recommended approach for encrypted data storage. Encrypted storage is based on WiredTiger, so all its advantages can be enjoyed using encryption as well. Encrypted storage is a part of MongoDB Enterprise Edition, the paid offering by MongoDB.

Using MongoDB's encrypted storage gives the advantage of increased performance versus encrypted storage volumes. MongoDB's encrypted storage has an overhead of around 15% compared to 25% or more for third-party encrypted storage solutions.

In most cases, if we need to use encrypted storage, we will know it well in advance from the application design phase, and we can perform benchmarks against different solutions to choose the one that best fits our use case.

Key management, replication, and auditing

MongoDB encryption at rest is only available in MongoDB Enterprise and MongoDB's cloud offering, MongoDB Atlas.

Key management can be performed either using a third-party server via KMIP or by storing a key file that's local to the DB server – one for each server.

Using the key file option does not meet most regulatory requirements and is strongly discouraged as it can become the prime attack vector against our system.

Using the KMIP option, we need to configure the following parameters to set up encryption at rest:

- `AuditLog.auditEncryptionKeyIdentifier`: The unique ID of the KMIP key that's used for audit log encryption.
- `AuditLog.compressionMode`: This compresses the audit log files. It defaults to none. If defined, it must go together with the previous option. `auditEncryptionKeyIdentifier`: At the time of writing, the only compression option is `zstd`.
- `auditEncryptionHeaderMetadataFile`: This specifies the path and filename for the header metadata, which we can use to decrypt the audit log file.
- `auditEncryptKeyWithKMIPGet`: Setting this to true will enable backward compatibility with KMIP servers running v1.0 or v.1.1 of the protocol. The latest version, 3.0, was released in 2021.

> **Note**
>
> While we should use KMIP for staging and production environments, we can also test audit log encryption locally for development by using the `auditLog.localAuditKeyFile` setting by using our keys in the same storage as the MongoDB server.

Encryption at rest using replica sets or shards is the same as when we use a standalone server. We need to set up encryption at rest in every single node, and as a security precaution, use different keys for each server.

Auditing using encryption at rest is also possible but only when using a KMIP server. In a mixed-use environment where some servers use encryption at rest and others don't, it's important to make sure that we don't leak **personally identifiable information** (**PII**) through the logs.

We can achieve this by setting the configuration parameter:

```
security.redactClientLogData=true
```

Audit logs cannot output to `syslog`. They can only output to a file.

Client-side field-level encryption

MongoDB introduced end-to-end client and field-level encryption with MongoDB 4.2.

This can be used in combination with TLS/SSL authentication between each server in the transport layer to avoid man-in-the-middle attacks. MongoDB storage with encryption at rest can also be used to avoid data being accessedfrom when it is stored in the disk.

Client-side field-level encryption encrypts data on the client side, making it impossible for even third-party systems that store our data (such as MongoDB Atlas) or employees working for our organization to read the secured underlying data that's been stored.

We can encrypt fields by using the automatic approach or by defining our explicit encryption of fields.

We will cover the automatic approach in more detail shortly. Our application needs to create a MongoClient object that conforms with the strict subset of the JSON Schema Draft 4 standard syntax to connect to the database (`https://datatracker.ietf.org/doc/html/draft-zyp-json-schema-04`).

So, data encryption and decryption are transparent to the client. Data encryption can be either deterministic or randomized.

Deterministic data encryption can be used for read queries, though doing so makes you susceptible to frequency analysis recovery attacks. For example, if we encode a boolean value, it would result in two encrypted values in the database. We only need to know the mapping between one of the DB encrypted values and its unencrypted equivalent to map it to true or false and the other value to false or true, respectively.

Randomized data encryption will overcome this problem by encrypting the entire object instead of the encrypted field. This resolves the frequency analysis attack problem that the per-field deterministic data encryption poses. However, this prevents us from querying the encrypted data effectively. Every query must now decrypt the entire object to compare the data with the query predicate.

We can and should use MongoDB's built-in JSON schema validation with encrypted fields to make sure that all the clients are submitting encrypted data to the server in the correct shape and format.

In-memory

Utilizing MongoDB's in-memory storage is a risky task with high rewards. Keeping data in memory can be up to 100,000 times faster than durable storage on disk.

Another advantage of using in-memory storage is that we can achieve predictable latency when we write or read data. Some use cases dictate latency that does not deviate from the norm, no matter what the operation is.

On the other hand, by keeping data in memory, we are open to power loss and application failures, which can result in us losing all of our data. Using a replica set can safeguard us against some classes of errors, but we will always be more exposed to data loss if we store it in memory as opposed to on disk.

For example, we can set up a replica set with two servers using in-memory storage and then set up a third server using WiredTiger storage with `{hidden: true, priority:0}`. This pair of settings guarantees that the WiredTiger node will never become primary but is otherwise continuously synced with the in-memory nodes.

However, there are some use cases in which we may not care that much about losing older data. For example, in the financial world, we may have the following:

- High-frequency trading/algorithmic trading, where higher latency in the case of high traffic can lead to transactions not being fulfilled.

- In fraud detection systems, we are concerned about real-time detection being as fast as possible and we can safely store only the cases that require further investigation, or the positive ones, in durable storage.

- Credit card authorizations, trade ordering reconciliation, and other high-traffic systems that demand a real-time answer.

In the web applications ecosystem, we have the following:

- In intrusion detection systems, such as fraud detection, we are concerned with detecting intrusion as fast as possible without so much concern for false positive cases

- In the case of the product search cache, losing data is not mission-critical, but rather a small inconvenience from the customer's perspective

- For real-time personalized product recommendations, there is a low-risk operation in terms of data loss. We can always rebuild the index, even if we suffer data loss

A major disadvantage of an in-memory storage engine is that our dataset has to fit in memory. This means we must know and keep track of our data usage so that we don't exceed the memory of our server.

At the time of writing, MongoDB 5.x requires the primary server to be using the WiredTiger storage engine if we want to use transactions in a replica set or sharded environment.

Overall, using MongoDB's in-memory storage engine may be useful in some edge use cases, but lacking durability in a database system can be a blocking factor in terms of its adoption.

> **Note**
> In-memory storage is part of MongoDB Enterprise Edition, the paid offering by MongoDB.

Other storage engines

The modular MongoDB architecture allows third parties to develop storage engines.

RocksDB and TokuMX

RocksDB is an embedded database for key-value data. It's a fork of `LevelDB` that stores key-value pairs in arbitrary byte arrays. It was started by Facebook in 2012 and now serves as the backend for the interestingly named **CockroachDB**, the open source database inspired by Google Spanner.

Another widely used MongoDB storage engine was TokuMX by Percona. TokuMX was designed with both MySQL and MongoDB in mind, but since 2016, Percona has focused its efforts on the MySQL version, instead switching over to **RocksDB** for MongoDB storage support.

MongoRocks is a project backed by Percona and Facebook that aims to bring the RocksDB backend to MongoDB. RocksDB can achieve higher performance than WiredTiger for some workloads and is worth investigating.

MongoRocks was deprecated by Percona in 2018 in favor of WiredTiger and has since been substituted by Percona Server for MongoDB.

Percona Server for MongoDB

Percona Server for MongoDB 5.0 is a drop-in replacement for MongoDB Community Edition. It offers some of the MongoDB Enterprise Edition features for free as source-available software.

Percona Server for MongoDB 5.0 uses the Creative Commons Attribution 4.0 International License. This license allows you to share, copy, and redistribute the material in any medium or format. The license allows you to adapt, remix, transform, and build upon the code for any purpose, even commercial. The only restriction is that attribution and credit must be given to Percona, and you must also provide a link to the license and an indication that you have made changes to the source code.

Licensing aside, Percona Server for MongoDB 5.0 adds the following free additions to MongoDB Enterprise Edition:

- LDAP authentication (simple LDAP authentication and external SASL authentication)
- Kerberos authentication
- Encryption at rest using Hashicorp vault as the key server
- Percona's in-memory storage engine
- LDAP authorization
- Audit logging
- Log redaction
- Replica set hot backup functionality
- SNMP monitoring

Percona Server for MongoDB 5.0 is based on MongoDB Community Edition 5.0 and as such will not benefit from any features being added to snapshot versions 5.1, 5.2, or 5.3. These snapshot versions are immediately available in MongoDB Atlas, the DBaaS offering from MongoDB Inc.

In this section, we discussed pluggable storage engines and how they can be used to improve performance and consistency with MongoDB. The next section will focus on the database locking concept and all we need to know before choosing or modifying our storage engine.

Locking in MongoDB

Document-level and collection-level locking is mentioned throughout this chapter and also in several other chapters in this book. It is important to understand how locking works and why it is important.

Database systems use the concept of locks to achieve ACID properties. When there are multiple read or write requests coming in parallel, we need to lock our data so that all the readers and writers have consistent and predictable results.

MongoDB uses multi-granularity locking. The available granularity levels, in descending order, are as follows:

- Global
- Database
- Collection
- Document

The locks that MongoDB and other databases use are as follows, in order of granularity:

- *IS*: Intent shared
- *IX*: Intent exclusive
- *S*: Shared
- *X*: Exclusive

If we use locking at a granularity level with *S* or *X* locks, then all the higher levels need to be locked with an intent lock of the same type.

Other rules for locks are as follows:

- A single database can simultaneously be locked in *IS* and *IX* mode
- An exclusive (*X*) lock cannot coexist with any other lock
- A shared (*S*) lock can only coexist with *IS* locks

Reads and writes requesting locks are generally queued in **first-in, first-out** (**FIFO**) order. The only optimization that MongoDB will do is reorder requests according to the next request in the queue to be serviced.

So, let's say we have an *IS(1)* request coming up next and our current queue is *IS(1)->IS(2)->X(3)->S(4)->IS(5)*, as shown here:

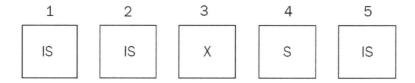

Figure 10.2 – The requests sequence before reordering

Here, MongoDB will reorder the requests as *IS(1)->IS(2)->S(4)->IS(5)->X(3)*, as shown here:

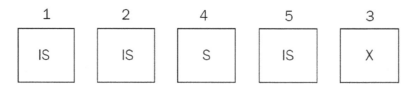

Figure 10.3 – The requests sequence after reordering

If, during servicing, the *IS(1)* request or new *IS* or *S* requests come in, such as *IS(6)* and *S(7)*, in that order, they will still be added at the end of the queue and won't be considered until the *X(3)* request has been dealt with.

Our new queue will now be *IS(2)->S(4)->IS(5)->X(3)->IS(6)->S(7)*:

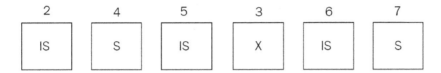

Figure 10.4 – The final requests sequence

This is done to prevent the starvation of the *X(3)* request, which would end up getting pushed back in the queue all the time due to new *IS* and *S* requests coming in. It is important to understand the difference between intent locks and locks themselves. The WiredTiger storage engine will only use intent locks for global, database, and collection levels.

It uses intent locks at higher levels (that is, collection, database, and global) when a new request comes in and according to the following compatibility matrix:

Mode	NL	IS	IX	S	SIX	X
NL	Yes	Yes	Yes	Yes	Yes	Yes
IS	Yes	Yes	Yes	Yes	Yes	No
IX	Yes	Yes	Yes	No	No	No
S	Yes	Yes	No	Yes	No	No
SIX	Yes	Yes	No	No	No	No
X	Yes	No	No	No	No	No

Figure 10.5 – Intent locks compatibility matrix

First, MongoDB will acquire intention locks in all ancestors before acquiring the lock on the document itself. This way, when a new request comes in, it can quickly identify if it cannot be serviced based on less granular locks.

WiredTiger will use *S* and *X* locks at the document level. The only exception to that is for typically infrequent and/or short-lived operations involving multiple databases. These will still require a global lock, similar to the behavior MongoDB had in pre 2.x versions.

> **Note**
>
> Administrative operations, such as dropping a collection, still require an exclusive database lock.

MMAPv1, as explained previously, uses collection-level locks. Operations that span a single collection but may or may not be spanning a single document will still lock up the entire collection. This is the main reason why WiredTiger is the preferred storage solution for all new deployments.

Lock reporting

We can inspect the lock status using any of the following tools and commands:

- `db.serverStatus()` through the `locks` document
- `db.currentOp()` through the `locks` field
- `mongotop`
- `mongostat`
- MongoDB Cloud Manager
- MongoDB Ops Manager

Lock contention is a really important metric to keep track of as it can bring our database to its knees if it gets out of control.

> **Note**
>
> If we want to terminate an operation, we must use the `db.killOp()` shell command.

Lock yield

A database with a database-level lock will not be really useful under stress and will end up being locked up most of the time. A smart solution to this in the early versions of MongoDB was getting operations to yield their locks based on some heuristics.

Any `update()` commands that affect multiple documents would yield their X lock to improve concurrency.

MMAPv1's predecessor in earlier versions of MongoDB would use these heuristics to predict whether data was already in memory before performing the requested operation. If it wasn't, it would yield the lock until the underlying operating system pages data in memory, and then re-acquire the lock to continue servicing the request.

The most notable exceptions to these are index scans, where the operation will not yield its lock and will just block on waiting for the data to get loaded from disk.

Since WiredTiger only uses intent locks at the collection level and above, it doesn't need these heuristics. This is because intent locks don't block other readers and writers.

Commonly used commands and locks

The most commonly used commands and locks are shown in the following table:

Command	Lock
`find()`	S
`it()` (query cursor)	S
`insert()`	X
`remove()`	X
`update()`	X
`mapreduce()`	Both S and X, depending on the use case. Some MapReduce chunks can run in parallel.
`index()`	Foreground indexing: Database lock. Background indexing: No lock, except for administrative commands that will return an error. Also, background indexing will take considerably more time.
`aggregate()`	S

Table 10.2 – Commands and locks for performing locking

Commands requiring a database lock

The following commands require a database lock. We should plan before issuing them in a production environment:

- `db.collection.createIndex()` in the (default) foreground mode
- `reIndex`
- `compact`
- `db.repairDatabase()`
- `db.createCollection()` if creating a multiple GB capped collection
- `db.collection.validate()`
- `db.copyDatabase()`, which may lock more than one database

We also have some commands that lock the entire database for a short period:

- `db.collection.dropIndex()`
- `db.getLastError()`

- `db.isPrimary()`
- Any `rs.status()` command
- `db.serverStatus()`
- `db.auth()`
- `db.addUser()`

These commands shouldn't take more than a few milliseconds to operate, so we shouldn't worry about this unless we have automated scripts with these commands in place, in which case we must note how often they will occur.

> **Note**
> In a sharded environment, each `mongod` applies locks, thus greatly improving concurrency.

In replica sets, our primary server must take all write operations. For these to be replicated correctly to the secondaries, we must lock the local database that holds the oplog of operations at the same time that we lock our primary document/collection/database. This is usually a short-lived lock that, again, we shouldn't worry about.

Secondaries in replica sets will fetch write operations from the primary local database's oplog, apply the appropriate X lock, and apply service reads once the X locks are done with.

From the preceding explanation, it's evident that locking should be avoided at all costs in MongoDB. We should design our database so that we avoid as many X locks as possible, and when we need to take X locks over one or multiple databases, we should do so in a maintenance window with a backup plan in case operations take longer than expected.

Again, WiredTiger only uses intent locking at the document level and as such has improved dramatically in terms of locking the database up. This section should mostly serve as a reference for troubleshooting legacy deployments using the now deprecated MMAPv1 storage engine.

Further reading

Please refer to the following links for further reference:

- MongoDB concurrency: `https://docs.mongodb.com/manual/faq/concurrency/`
- MongoDB storage engines: `https://docs.mongodb.com/manual/core/storage-engines/`
- MongoDB pluggable storage engines blog post: `https://www.mongodb.com/blog/post/building-applications-with-mongodbs-pluggable-storage-engines-part-2`

- MongoDB WiredTiger: `https://docs.mongodb.com/manual/core/wiredtiger/`

- MongoDB's create index and collection options: `https://docs.mongodb.com/manual/reference/method/db.collection.createIndex/#createindex-options`

- `https://docs.mongodb.com/manual/reference/method/db.createCollection/#create-collection-storage-engine-options`

- MongoDB deprecated MMAPv1 storage engine documentation: `https://docs.mongodb.com/manual/core/mmapv1/`

- MongoDB WiredTiger WT Session documentation: `http://source.wiredtiger.com/mongodb-6.0/struct_w_t___s_e_s_s_i_o_n.html`

- MongoDB with the microservices architecture: `https://www.mongodb.com/what-are-microservices`

- MongoDB storage engines overview: `https://webassets.mongodb.com/storage_engines_adress_wide_range_of_use_cases.pdf`

- MongoDB read concern: `https://docs.mongodb.com/manual/reference/read-concern/`

Summary

In this chapter, we learned about different storage engines in MongoDB. We identified the pros and cons of each and the use cases for choosing each storage engine.

Then, we learned about using multiple storage engines, how we can use them, and the benefits. A big part of this chapter was also dedicated to database locking, how it can happen, why it is bad, and how we can avoid it.

We split our operations by the lock they need. This way, when we design and implement our application, we can make sure that we have a design that locks our database as little as possible.

In the next chapter, we will learn about MongoDB and how we can use its tools on premises or in the cloud to design and develop our applications.

11
MongoDB Tooling

Features, stability, and great driver support are all important; however, there is another area that is key to a software product succeeding, and that is the ecosystem built around it. MongoDB (originally under the name 10gen Inc.) introduced the **MongoDB Management Service (MMS)** more than 10 years ago, in 2011, and it was seen as an innovation at the time. In this chapter, we will go through the following suite of different tools that are available for MongoDB and explore how they can improve productivity:

- MongoDB Charts
- MongoDB Kubernetes Operator
- MongoDB Atlas Serverless
- MongoDB Realm services
- MongoDB Compass
- MongoDB Cloud and Ops Manager

In particular, we will cover the following topics:

- Introduction to MongoDB tools
- MongoDB Kubernetes Operator
- MongoDB Atlas Serverless
- MongoDB Realm
- MongoDB Charts
- MongoDB Compass
- MongoDB Cloud Manager

Technical requirements

You need to have access to `https://www.mongodb.com/` and sign up for a trial account using the website. A free account is enough for most of the examples demonstrated in this chapter, but any paid account is needed for a few of the most advanced examples.

Introduction to MongoDB tools

MongoDB tools fall into two categories: on-premises and cloud-based services. Over time, **software-as-a-service (SaaS)** tooling has prevailed over on-premises tooling. While a few tools are still available as a standalone download, MongoDB has shifted efforts toward cloud-based tools around their online platform, MongoDB Atlas.

Readers of previous editions of this book will identify the shift in focus throughout the chapters. New readers should be aware that while tools can be available on-premises, there is a possibility that this won't be the case in the future.

The MongoDB tool we will cover first is the MongoDB Atlas platform.

MongoDB Atlas

MongoDB Atlas is MongoDB's cross-cloud platform offering. It offers the latest stable version of MongoDB, along with an ecosystem of tooling from the creators, MongoDB Inc. In the next section, we will learn how to initiate a replica set MongoDB cluster in any of the major cloud providers (**Amazon Web Services (AWS)**, Azure, or **Google Cloud Platform (GCP)**).

Setting up a cluster

MongoDB Atlas is MongoDB's **database-as-a-service (DBaaS)** offering. It is available as a multi-cloud offering, supporting AWS, Microsoft Azure, and GCP.

Using DBaaS, patches and minor version upgrades are applied automatically, without any downtime. Using the **graphical user interface (GUI)**, a developer can deploy geographically distributed database instances to avoid having any **single point of failure (SPOF)**. For websites with significant traffic, this can also help by placing database servers closer to the users who are accessing their data. This is a key part of MongoDB's strategy and offering as they embrace having data close to the user.

Similar to most DBaaS offerings, Atlas allows users to scale deployment using a GUI. Every deployment lives on its own **virtual private cloud (VPC)** and can leverage MongoDB Enterprise Server features such as encryption key management, **Lightweight Directory Access Protocol (LDAP)**, and auditing features.

The live migration service can be used to migrate datasets from existing deployments on-premises or even relational databases such as Oracle, Microsoft SQL Server, PostgreSQL, and MySQL, as of version 6, with the relational migration tool, in any of the three supported cloud providers or other DBaaS services, such as **Compose** and **ObjectRocket**, using the same GUI.

We can query our datasets in MongoDB Atlas using **MongoDB Query Language** (**MQL**) or the **Structured Query Language** (**SQL**) interface for Atlas. The SQL interface for Atlas uses a SQL-92-compatible dialect that is designed from the ground up for the document model. The Atlas SQL interface can also query across different Atlas clusters and heterogeneous cloud storage such as **Simple Storage Service** (**S3**) buckets.

Creating a new cluster

Creating a new cluster using MongoDB Atlas is as simple as clicking and selecting configuration options. In the following screenshot, we can see all the options that are available when creating a new cluster:

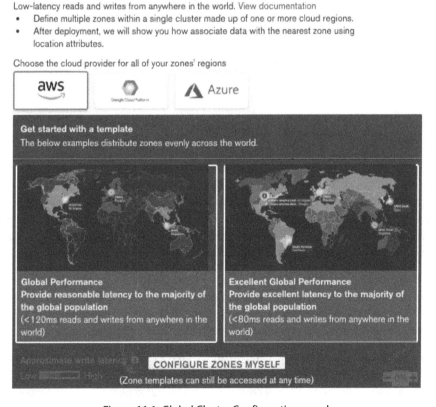

Figure 11.1: Global Cluster Configuration panel

The following screenshot shows the **Zone configuration summary** panel:

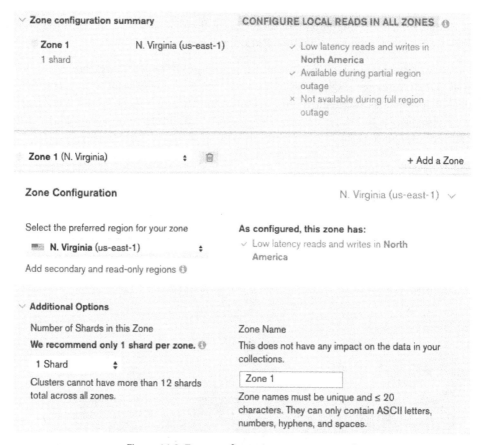

Figure 11.2: Zone configuration summary panel

One of the game-changing features in MongoDB Atlas is the ability to instantly provision geographically distributed servers across different zones and data centers (for all three major cloud providers) with the goal of having our data closest to our users. This can be useful for performance and regulatory reasons (such as **General Data Protection Regulation (GDPR)**, for the **European Union (EU)**).

By enabling global writes, we can start configuring this setting. Using any of the two templates—**Global Performance** or **Excellent Global Performance**—an administrator can create server configurations that will be less than 120 milliseconds or 80 milliseconds, respectively from any user around the world. An administrator can also define their own custom assignments from regions to data centers.

In the **Zone configuration summary** panel, we can see an overview of how our settings will affect performance. **M30** is the shard-enabled MongoDB Atlas plan, and this configuration is creating (under the hood) a shard per zone. We can create more shards per zone but this is not recommended at this time.

Enabling **CONFIGURE LOCAL READS IN ALL ZONES** will create local read-only replica set nodes in every region other than the zone that is used to write data to. So, if we have three zones (*A*, *B*, and *C*), we end up with writes for *A* going to *A* but reads from *A* happening from a server in either zone *A*, *B*, or *C*, depending on which server is geographically closer to the user. The same goes for zones *B* and *C*.

This section is probably the most important for complex, multi-region deployments and should be treated with extreme care.

The next section is for configuring the servers that we want to use for our clusters, as illustrated in the following screenshot:

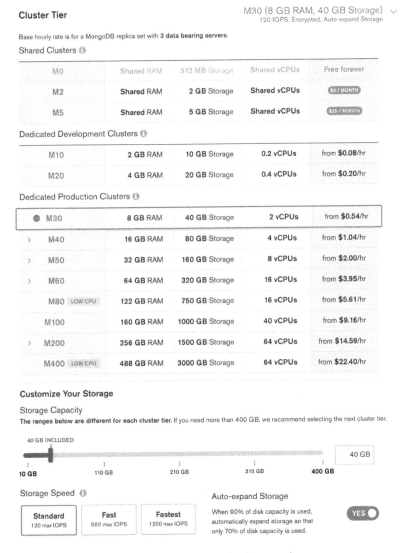

Figure 11.3: Server selection panel

This is similar to how we select servers in **Elastic Compute Cloud (EC2)** or Microsoft Azure. The main point to note is that we can select custom **input/output operations per second (IOPS)** performance and that we should select the **Auto-expand Storage** option to avoid running out of disk capacity. Together with this option, it's always useful to keep an eye on storage allocation to avoid excessive charges at the end of a billing cycle.

In the next panel, we can configure backup and advanced options for our cluster. The following screenshot shows the **Additional Settings** panel for continuous backup:

Additional Settings

MongoDB 4.0, Backup ∨
Continuous Backup

Select a Version

4.0 with WiredTiger™ ⬍

All clusters launch with the WiredTiger™ storage engine.

Turn on Backup **(M10 and up)**

YES●

You can easily enable or disable backups at any time after deploying.
Compare Backup Solutions

Continuous
Point-in-time data recovery and fast, granular data restores

- **$1.50 - $2.50/GB Month***
- Restore to any point in time within the last 24 hours
- Quickly restore with queryable snapshots
- Backups are stored in the same country** as your database
- 12 months of snapshots by default

Cloud Provider Snapshots
Localized backup storage with fast restore times for
snapshot images

- Starting at $0.19/GB month*
- Fast time to restore
- Backups are stored in the same cloud region as your database
- Snapshots are taken at 24 hour increments
- 3 Snapshots stored by default

* Charges will vary depending on database size and backup retention schedule.
** For the US, UK, Ireland, Germany, and Australia. Learn more

* Total charges will vary depending on the region you choose, the size of each snapshot and the number of snapshots retained. Learn more

Figure 11.4: Additional Settings panel

The following screenshot shows the **Advanced Settings** option for enabling **BI Connector**:

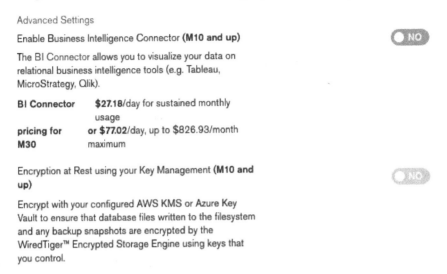

Figure 11.5: Advanced Settings panel

The following screenshot shows the **More Configuration Options** panel, with options that are available:

Figure 11.6: More Configuration Options panel

Using the steps outlined previously, we can create a new MongoDB replica set cluster using a flexible configuration.

Important notes

MongoDB has provided some useful tips regarding MongoDB Atlas, including the following:

- Use the latest version of MongoDB whenever possible.

- Use the latest **Transport Layer Security** (**TLS**) version available, which is 1.3 at the time of writing.

- Encryption at rest cannot be used together with continuous backups. We need to select cloud provider snapshots to be able to use this feature.

- It's probably a good idea to disable server-side JavaScript unless we know why we need it—for example, when we have legacy MapReduce jobs in place.

- Requiring indexes for all queries can be useful if we have a well-defined business case and requirements for how we will use the database, and/or we expect our dataset to be so large that querying without indexes is practically impossible.

- Finally, we get to select the name of our cluster. This cannot be changed after creation, so it's important to agree with fellow team members before we click on the **Create Cluster** button.

After a bit of waiting, our cluster will be up and running, and we will be able to connect to it via the MongoDB **Uniform Resource Identifier** (**URI**). In the next section, we will learn how to search our data for keywords and patterns.

MongoDB Atlas Search

MongoDB Atlas offers advanced search capabilities based on Apache Lucene (https://lucene.apache.org/) technology. This is separate and independent from the legacy text search capability that we can use in a self-managed environment with the $text operator.

MongoDB Atlas Search is much more advanced and performant than the legacy text search and should be preferred where possible.

Creating and searching using Atlas indexes

First of all, we need to create indexes that Atlas Search will use to query our database. We can do this by going to our database page and selecting **Search** from the options at the top of the page.

We can create indexes using the visual editor or the **JavaScript Object Notation** (**JSON**) editor. The visual editor guides the user through steps to create a JSON configuration file.

Depending on our use case, we have to choose between dynamic and static field mapping.

Dynamic mapping will index every data type and cross all documents in a collection. This is useful when we have unstructured data for which we can't know the schema **ahead of time** (**AOT**).

Static mapping will only index the fields that we define in the JSON configuration file. We can define top-level fields or nested fields using the `.` notation, such as `address.name`.

We can store the full document in the index itself and return it with the query result without querying the underlying database collection. This will avoid an extra query in the database collection but can put a strain on the index size and performance.

To avoid querying the database collection, we need to set `returnStoredSource` to `true` at the time of querying.

When we index a string field, we can use one of the built-in index analyzers or define our own. An index analyzer will transform the data field value to the term that will be stored in the index.

The built-in analyzers are listed here:

- **Standard** (default)
- **Simple**: Will break input text into tokens separated by any non-letter character.
- **Whitespace**: Will break input text into tokens separated by any whitespace character.
- **Language**: Will transform input text using language-specific text analyzers.
- **Keyword**: Will index input terms as a whole. This is useful, for example, when we need to perform matching only on a whole phrase.

We can also define a custom analyzer if the built-in ones don't fit our needs. Our custom analyzer must define a tokenizer that will break the input text into tokens. The custom analyzer may define a character or token filter to filter input tokens through and transform them according to the filter.

Similar to the index analyzer, a search analyzer will transform our query to the searchable term that Atlas Search will try to match against the index. Faceted search allows developers to implement narrowing down search results by applying filters based on the classification of items. An example of facets could be searching for a product on a site such as Amazon and then drilling down on results based on category (string), price range (numeric), or prime availability (Boolean). MongoDB supports faceted search using the `$facet` operator in the aggregation pipeline. While this is fully functional, it does not perform adequately as the number of documents grows. MongoDB Atlas supports faceted search using the underlying Apache Lucene engine, which performs better (up to 100x faster than `$facet`), and can scale almost linearly with the input size. We can use a facet on **String**, **Number**, and **Date** valued fields.

In this section, we learned about Atlas and Atlas Search functionality. In the next section, we will go through how Kubernetes and MongoDB work together.

MongoDB Kubernetes Operator

In this section, we will learn how to use Kubernetes with MongoDB deployments.

An introduction to Kubernetes

Kubernetes (`https://kubernetes.io`) is an open source container orchestration system for automating deployment, scaling, and the management of containerized applications. In layman's terms, we can use Kubernetes (often referred to as *K8s*) to manage applications deployed via containers. Kubernetes was initially developed at Google and is now maintained by the **Cloud Native Computing Foundation** (**CNCF**).

The most widely used container technology is probably Docker. We can download and install Docker on any PC and, through a few commands, install a Docker image that will be isolated from our host system and contain our application code. Docker performs operating system-level virtualization, where all containers are run by the host's operating system kernel. This results in containers being more lightweight than a full **virtual machine** (**VM**).

Multiple Docker containers can be orchestrated using **Docker Swarm**. This is similar to Kubernetes, and sometimes the two systems are directly compared with each other.

MongoDB provides tools that can help administrators deploy and manage MongoDB clusters using Kubernetes.

Enterprise Kubernetes Operator

Starting from MongoDB 4.0, the MongoDB Enterprise Kubernetes Operator enables a user to deploy and manage MongoDB clusters directly from the Kubernetes **application programming interface** (**API**). This circumvents the need to directly connect to Cloud Manager or Ops Manager and simplifies the deployment and management of Kubernetes clusters.

> **Note**
> Cloud Manager is, in most aspects, the SaaS equivalent of Ops Manager.

Enterprise Kubernetes Operator can be installed using Helm, the package manager for Kubernetes. First, we have to clone the GitHub repository from MongoDB, located at `https://github.com/mongodb/mongodb-enterprise-kubernetes.git`.

After we change the directory to our local copy, we can issue the following command:

```
helm install helm_chart/ --name mongodb-enterprise
```

We will then have the local copy installed; the next step is to configure it.

By configuring our local installation, we need to apply a Kubernetes `ConfigMap` file. The configuration settings that we need to copy from Ops Manager or Cloud Manager are listed here:

- **Base URL**: The **Uniform Resource Locator** (**URL**) of your Ops Manager or Cloud Manager service. For Cloud Manager, this will be `http://cloud.mongodb.com`; for Ops Manager, this should be similar to `http://<MY_SERVER_NAME>:8080/`.

- **Project ID**: The **identifier** (**ID**) of an Ops Manager project that the Enterprise Kubernetes Operator service will deploy into. This should be created within the Ops Manager or Cloud Manager service and is a **unique ID** (**UID**) to organize a MongoDB cluster and provide a security boundary for the project. It should be a 24-digit hexadecimal string.

- **User**: An existing Ops Manager username. This is the email of a user in Ops Manager that we want the Enterprise Kubernetes Operator service to use when connecting to Ops Manager.

- **Public API key**: This is used by the Enterprise Kubernetes Operator service to connect to the Ops Manager **REpresentational State Transfer** (**REST**) **API** endpoint.

A public API is created by clicking on the username on the **Ops Manager** console and selecting **Account**. On the next screen, we can click on **Public API Access**, and then click on the **Generate key** button and provide a description. The next screen will display the public API key that we need.

> **Note**
> This is the only chance that we will ever have to view this API key, so we need to write it down; otherwise, we will need to regenerate a new key.

Once we have these values, we can create a Kubernetes `ConfigMap` file with any name we want, as long as it's a `.yaml` file. In our case, we will name it `mongodb-project.yaml`.

Its structure will look like this:

```
apiVersion: v1
kind: ConfigMap
metadata:
name:<<any sample name we choose(1)>>
namespace: mongodb
data:
projectId:<<Project ID from above>>
baseUrl: <<BaseURI from above>>
```

Then, we can apply this file to Kubernetes using the following command:

```
kubectl apply -f mongodb-project.yaml
```

The last step we need to take is to create a Kubernetes Secret. This can be done using the following command:

```
kubectl -n mongodb create secret generic <<any sample name for
credentials we choos>> --from-literal="user=<<User as above>>"
--from-literal="publicApiKey=<<our public api key as above>>"
```

> **Note**
> We need to note down the credential's name as we will need it in the subsequent steps.

Now, we are ready to deploy our replica set using Kubernetes! We can create a `replica-set.yaml` file with the following structure:

```
apiVersion: mongodb.com/v1
kind: MongoDbReplicaSet
metadata:
name: <<any replica set name we choose>>
namespace: mongodb
spec:
members: 3
version: 3.6.5
persistent: false
project: <<the name value (1) that we chose in metadata.name of
ConfigMap file above>>
credentials: <<the name of credentials secret that we chose
above>>
```

We apply the new configuration using `kubectl apply`, as follows:

```
kubectl apply -f replica-set.yaml
```

We will be able to see our new replica set in Ops Manager.

> **Note**
> To troubleshoot and identify issues in MongoDB using Kubernetes, we can use `kubectl logs` to inspect logs, and `kubectl exec` to shell into one of the containers that are running MongoDB.

In this section, we learned how to use Kubernetes with MongoDB. The next section will show why and when to use serverless instances in MongoDB Atlas.

MongoDB Atlas Serverless

Serverless has been quickly gaining traction over the past few years. Serverless computing doesn't mean that there is no server anywhere—it means that the server infrastructure is completely transparent to the application developer. The cloud infrastructure abstracts any required servers away from the developer.

The developer can write code that executes directly and can access storage and computing resources without the need to think as much about scaling or storage needs.

Connecting to a serverless instance is the same as connecting to a MongoDB Atlas replica set, such as the MongoDB URL.

Serverless computing's major advantage is that we only pay for the resources that we are using. This can result in cost savings when we have a highly variable load, a load with short bursts, or otherwise cannot fully utilize our servers for most of the time.

Serverless instances are in preview mode as of the time of writing this book (summer 2022) and have some limitations. The limitations are gradually removed but we should check the documentation if we are considering using serverless deployment in Atlas.

Serverless instances require up-to-date versions for application drivers.

Serverless is typically better suited when we want to get started with minimal effort or we have a workload with bursts in traffic that require quick scaling up and down. We can also use serverless if we want to quickly test a workload in the MongoDB Atlas environment.

During the MongoDB Atlas Serverless preview release, we should not use it for production workloads. The next section will introduce us to MongoDB Realm, the integrated environment for cross-platform application development from MongoDB.

MongoDB Realm

MongoDB Realm substitutes and supersedes MongoDB Mobile, Stitch, Sync, and other earlier MongoDB application development services. Realm offers an integrated environment to develop MongoDB-based real-time, offline-first applications across heterogeneous platforms.

Realm offers a mobile database that can run on Android or iOS and can offline/online synchronize with a MongoDB Atlas instance in the backend.

MongoDB's Realm offering aims to make development faster and easier by abstracting away backend development and common services from application developers.

The following diagram provides an overview of the Realm services stack:

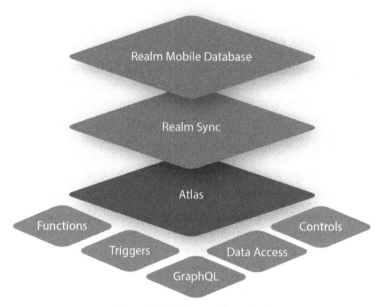

Figure 11.7: Realm services stack

The layered approach in *Figure 11.7* shows how a mobile database can interact with the backend Atlas database using the Realm Sync service.

Realm Sync

Realm Sync can abstract a MongoDB Atlas database in a mobile application on an iOS or Android platform. The mobile application uses an embedded database in the device that transparently synchronizes with the backend MongoDB Atlas instance.

Realm Sync is built with the assumption that connectivity to the Atlas instance is intermittent. Data is always saved locally and synced to the server whenever the connection is restored.

Any updates to the local object will be propagated to the backend using a deterministic conflict resolution algorithm. Even if the changes are received by the backend out of their execution order, they will converge to the same final state in the backend and any other subscribing clients. This results in Realm Sync being characterized as a strongly eventual consistent system.

Finally, Realm Sync offers the capability to subscribe to document (object) updates. The concept of live objects allows developers to make sure that the object is always the most up-to-date copy, including any updates that may have occurred in the local mobile client database or the remote MongoDB Atlas database itself.

Realm Sync data model

The data model allows Realm Sync to propagate changes back and forth between the local mobile database and the MongoDB Atlas backend service. We can generate object models using a document schema or let Realm automatically generate language-specific schemas.

Sync object models from a software development kit (SDK) in development mode will automatically generate a document schema and collection in the MongoDB Atlas database. This approach is preferred when we need a client-first approach with custom data modeling baked into our code.

Generating object models from a document schema will automatically generate language-specific object models based on MongoDB Atlas-defined schemas. In a sense, this is the opposite approach: modeling our database first and generating client models out of it. We can generate language-specific models using the MongoDB Atlas Realm Web **user interface** (**UI**) or the Realm **command-line interface** (**CLI**) in a variety of languages including JavaScript, Kotlin, Java, and others.

Realm Sync mode

The next step after defining a data model is to define the Sync mode between **partition-based** and **flexible sync**.

Partition-based sync will partition data in each client database based on a key that naturally exists in the document (such as `user_id`) or a key that only exists to serve the partitioning purpose (such as `_partition`). The partition key will segment data values across clients.

Flexible sync allows us to define a query on the client side and dynamically sync the objects that match the query. This method is way more powerful than partition-based as we can sync and match and download only the required documents on each mobile client.

Realm data access rules

Partition-based sync data access can be configured based on the partition strategy and read/write permissions that we can define using a JSON schema.

Flexible sync data access rules are configured per collection using session roles and rules to determine which matching objects from the query can be accessed by the client user.

Enabling Realm Sync and best practices

The last step is enabling Realm Sync between the client(s) and the MongoDB Atlas backend, which can be done via the web UI, the CLI, or the API.

Some best practices for Realm Sync are noted here:

- Enable >=48 hours of **operations log** (**oplog**) data in the MongoDB Atlas cluster.
- Avoid breaking changes after enabling sync.

- Configure maximum offline time for each client. If this time period elapses, the client must perform a full data sync.

- Keep the MongoDB data source and the Realm application within the same cloud provider and geographic region.

Realm Sync acts as a bridge between mobile services and the MongoDB Atlas backend infrastructure. We can also use Realm independently with MongoDB Atlas backing services, which is a **backend-as-a-service (BaaS)** offering. We will learn more about this in the next section.

Realm Application Services

Realm Application Services is MongoDB's integrated offering for rapid application development. It is based on a serverless environment and offers a suite of services that allow developers to focus on application code instead of backend infrastructure.

We can use one of the existing Realm template apps to get started (`https://www.mongodb.com/docs/atlas/app-services/template-apps/`) or create our own (`https://www.mongodb.com/docs/atlas/app-services/manage-apps/create/create-with-realm-ui/#std-label-create-a-realm-app`) from the Realm UI, CLI, or API.

Next, we are going to learn more about the services that bridge application and backend development.

Realm functions

Realm functions are blocks of server-side JavaScript code that execute in the user caller's context, without the need to stand up a server (serverless model). Functions are typically used for lightweight, time-constrained processing. They can run for up to 120 seconds, using up to 256 **megabytes (MB)** of memory, open up to five sockets, and process up to 18 MB of input data into the function. In essence, Realm functions are functionally and conceptually similar to AWS Lambda functions.

Realm triggers

Realm triggers are similar to relational database triggers. Each trigger is linked to an event and triggers the execution of a **Realm function**. An event can be a database event such as a document insert, delete, or update operation. An event can also be an authentication type event such as a user login or logout. Finally, an event can be scheduled AOT, setting—for example—a daily recurrence.

Realm GraphQL

Realm allows users to query underlying data using GraphQL. We need to define a JSON schema for each collection that Realm uses to generate a GraphQL API definition, types, and resolvers. Setting up and accessing data using GraphQL follows the standard, and we can connect using the CLI or any compliant framework such as React-based Apollo.

User authentication

MongoDB Realm supports user authentication out of the box using built-in authentication providers and custom ones.

We can authenticate users using email/password, API keys, **Open Authorization 2.0 (OAuth 2.0)** using an Apple, Google, or Facebook login, and—finally—anonymous/no authentication.

We can roll our own authentication business logic using a custom **JSON Web Token (JWT)** or custom function.

Schemas

GraphQL, Realm Sync, and data access rules all use an App Services **Binary JSON (BSON)** schema to understand the shape of the data that they need to access or transform.

The BSON schema is an extension of the JSON schema (`https://json-schema.org/`), which is widely used to annotate and validate JSON documents.

The JSON schema defines the types of document fields, as in the following example:

```
{
    "title": "trade",
    "required": [
        "_id",
        "consigner",
        "consignee",
        "commodity_code",
        "price",
        "description"
    ],
    "properties": {
        "_id": { "bsonType": "objectId" },
        "consigner": { "bsonType": "string" },
        "consignee": { "bsonType": "string" },
        "commodity_code": { "bsonType": "string" },
        "price": { "bsonType": "number" },
    "description": {"bsonType":"string"}
    }
}
```

The preceding example would pass validation for the following document:

```
{
    "_id": ObjectId("5af712eff26b29dc5c51c60f"),
    "consigner": "British-exporter-Alex",
    "consignee": "France-importer-Bob",
    "commodity_code": "HS123445",
    "price": 1337
 "description": "sale of a gimmick from UK to France"
}
```

The BSON schema supports any of the following data types:

- Object
- Array
- String
- Number
- Boolean
- **Universally unique ID (UUID)**
- `ObjectId`
- Binary data
- Mixed
- Set
- Dictionary

HTTPS endpoints

We can use **Hypertext Transfer Protocol Secure (HTTPS)** endpoints to execute serverless functions that are invoked using standard HTTPS requests. We can use `GET`, `POST`, `PUT`, `PATCH`, and `DELETE` methods to trigger the function execution. We can configure the endpoint to use an email/password, API key, or custom JWT authentication. An HTTPS endpoint executes a Realm function and, as such, can perform a variety of tasks including integrating with external services or calculating complex business logic.

Static file hosting

We can use MongoDB Realm to host static files. These will be accessible at the following URL:

```
<AppID>.mongodbstitch.com
```

This can be used to store static files used in our application. We can only store files up to 25 MB in size, up to 1 **gigabyte (GB)** storage, and 20,000 files per Realm application. The limits imposed mean that we shouldn't use static file hosting to build a storage service, but they are more than enough to host any static files that we need in our application.

Realm values

Realm values and secrets are similar to HashiCorp Vault's functionality (`https://www.vaultproject.io/`). Values can be resolved to a string, array, or JSON object. Values are accessible in the context of functions and rule expressions.

We can define different values per environment that are environment values, similar to Unix environment variables.

We can also define Secrets—secret values that are not exposed to users—and exported applications. We can access Secrets by either linking the Secret to a value or by linking to the Secret by name in the authentication provider and service configuration.

Realm Studio

Realm Studio is a cross-platform GUI tool that allows us to edit Realm database files. It can help us perform administration tasks and export schema as class definitions in C#, Java, JavaScript, Kotlin, Swift, and TypeScript. The source code is available on GitHub at `https://github.com/realm/realm-studio/releases`.

Realm application services can greatly simplify and speed up application development, enabling us to focus on application code, away from infrastructure setup and maintenance. In the next section, we will learn more about the available GUI-based tools for MongoDB administration and operations.

MongoDB Charts

MongoDB Charts is a tool to generate visualizations from MongoDB data. It enables non-technical people to query a MongoDB database using a GUI and share the results with colleagues.

MongoDB Charts can create a series of charts, including the following:

- A column and bar chart reference
- A line and area chart reference

- Grid charts, such as the following:

    - A heatmap reference

    - A scatter chart reference

- A donut chart reference

- Text charts: Number chart reference

As with Ops Manager, it's a standalone executable utilizing Docker that needs to be installed and managed on-premises.

> **Note**
>
> Use replica set secondaries for chart queries. Ideally, use a secondary, hidden, non-electable node as an analytics node in replica sets.

MongoDB Charts is a simple and powerful tool from MongoDB. MongoDB Compass is another tool used for querying and data analysis that we will discuss in the next section.

MongoDB Compass

MongoDB Compass is similar to MongoDB Charts, with less functionality around charting, but more heavily oriented toward running ad hoc queries and connecting to our database without the need for a CLI.

Compass provides the ability to query MongoDB through a GUI and visually construct queries. It can provide rich visualizations of result datasets, and help with constructing aggregation queries through a point-and-click interface.

Compass also provides visualizations for most administrative queries around query and index performance so that it can be used to monitor and troubleshoot clusters from a database administrator perspective. It exposes an API that can be used to import or develop plugins.

A useful feature for non-technical users is the ability to download a read-only edition so that we can limit access to non-destructive operations. There is also an isolated edition of this tool that can be used to restrict connections to a single chosen server. These requests will also be TLS-encrypted.

Compass is available for Windows, macOS X, Red Hat, and Ubuntu as an executable download. MongoDB Compass has a limited free edition, and the full feature set is available via a MongoDB subscription package.

MongoDB Connector for Business Intelligence

MongoDB Connector for Business Intelligence (**MongoDB Connector for BI**) is one of the most useful tools for non-developers. It is a part of the MongoDB Enterprise Advanced subscription and enables integration with BI tools using standard SQL queries.

It enables MongoDB integrations with enterprise tools such as Tableau, Qlik, Spotfire, Cognos, MicroStrategy, and SAP BusinessObjects.

It is available as an executable download for Amazon Linux, Debian, macOS X, Red Hat, SUSE, Ubuntu, and Windows platforms, and can work with both on-premises databases and MongoDB Atlas. Once installed and configured correctly, it can provide an **Open Database Connectivity** (**ODBC**) **data source name** (**DSN**) that most BI tools can use to connect to.

The next section will introduce us to the MongoDB monitoring service tools, Cloud Manager and Ops Manager.

MongoDB Cloud Manager

Formerly known as **MongoDB Management Service** (**MMS**) and before that as **MongoDB Monitoring Service** (**MMS**), Cloud Manager is a managed SaaS for on-premises MongoDB deployments.

Atlas as a DBaaS solution can provide an **end-to-end** (**E2E**) solution for database management. For many use cases, this may not be feasible. In this case, it may make sense to use some of the features in a pay-as-you-go fashion.

Cloud Manager has a limited free tier and several paid ones.

Here are some of the key features of Cloud Manager:

- Automated backups
- More than 100 database metrics and **key performance indicators** (**KPIs**) that can be used to track MongoDB's performance
- Customized alerts that integrate with third-party systems such as PagerDuty, email, and **Short Message Service** (**SMS**)
- A unified operations view, either via directly querying its JSON API or by integrating it with popular performance-tracking solutions such as New Relic (`https://www.newrelic.com`)

Premium plans also offer advice around performance and indexing. Cloud Manager's only requirement is to install the required agents in our application.

MongoDB Ops Manager

Ops Manager is, in many ways, different from Cloud Manager. In contrast to Cloud Manager, it is a downloadable executable for Windows Server, **Red Hat Enterprise Linux** (**RHEL**), or Ubuntu.

Following on from this, users need to install and manage the service within their own infrastructure.

Aside from this difference, Ops Manager can also help achieve similar goals to Cloud Manager, as follows:

- Monitoring more than 100 performance metrics
- Automation on installing and upgrading clusters; plus index maintenance can be achieved with zero downtime
- For continuous, incremental backups with **point-in-time recovery** (**PITR**)
- Query optimization
- Index suggestions

A sample Ops Manager topology is provided here:

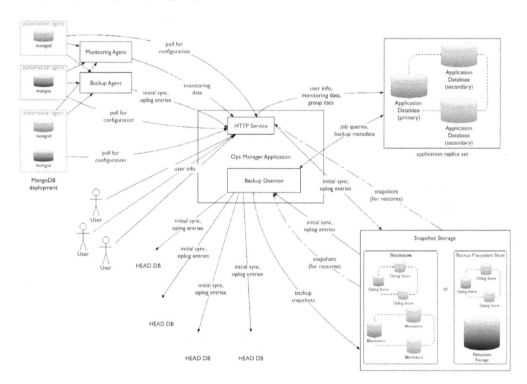

Figure 11.8: Ops Manager topology

Aside from the Ops Manager and the MongoDB nodes, we also need snapshot storage if we have enabled backups.

Ops Manager can be a better alternative to Cloud Manager if we want an on-premises solution for security or other reasons. This is included as part of the MongoDB Enterprise Server paid solution.

Summary

In this chapter, we went through different MongoDB tools and learned how to use them to increase productivity. Starting from MongoDB Atlas, the hosted DBaaS solution, we followed on with Atlas Search and Serverless. Then, we discussed Kubernetes, how it compares to Docker and Docker Swarm, and how we can use Kubernetes with MongoDB Enterprise Kubernetes Operator.

The next section was dedicated to MongoDB Realm for mobile, IoT, and web application development. We detailed the different offerings from MongoDB Atlas around data access, querying, triggers, serverless functions, authentication, hosting, secrets, schemas, values, and static file hosting.

Then, we dived into MongoDB Charts and MongoDB Compass—the GUI-driven MongoDB administration tools. We learned about MongoDB Connector for BI and how it can prove to be useful for our purposes. Finally, we learned about Cloud Manager and Ops Manager and explored how they differ from each other.

In the next chapter, we are going to switch gears and deal with how we can use big data with MongoDB to ingest and process large streaming and batch datasets.

12
Harnessing Big Data with MongoDB

MongoDB is often used in conjunction with big data pipelines because of its performance, flexibility, and lack of rigorous data schemas. This chapter will explore the big data landscape, and how MongoDB fits alongside message queuing, data warehousing, and **extract, transform, and load** (**ETL**) pipelines.

We will also learn what the MongoDB Atlas Data Lake platform is and how to use this cloud data warehousing offering from MongoDB.

These are the topics that we will discuss in this chapter:

- What is big data?
- Big data use case with servers on-premises
- MongoDB Atlas Data Lake

Technical requirements

To follow along with the examples in this chapter, we need to install Apache Hadoop and Apache Kafka and connect to a MongoDB cluster such as the one that we created using MongoDB Atlas in the previous chapters. We need a trial account for MongoDB Atlas to use MongoDB Atlas Data Lake.

What is big data?

In the last 10 years, the number of people accessing and using the internet has grown from a little under 2 billion to 3.7 billion to 5.3 billion as of 2022. Around two-thirds of the global population is now online.

With the number of internet users increasing, and with networks evolving, more data is being added to existing datasets each year. In 2016, global internet traffic was 1.2 **zettabytes** (**ZB**) (which is 1.2 billion **terabytes** (**TB**)) and has grown to an expected 3.6 ZB in 2022.

This enormous amount of data that is generated every year means that it is imperative that databases and data stores, in general, can scale and process our data efficiently.

The term *big data* was first coined in the 1980s by John Mashey (`http://static.usenix.org/event/usenix99/invited_talks/mashey.pdf`), and mostly came into play in the past decade with the explosive growth of the internet. Big data typically refers to datasets that are too large and complex to be processed by traditional data processing systems, and so need some kind of specialized system architecture to be processed.

Big data's defining characteristics are, in general, these:

- Volume
- Variety
- Velocity
- Veracity
- Variability

Variety and variability refer to the fact that our data comes in different forms and our datasets have internal inconsistencies. These need to be smoothed out by a data cleansing and normalization system before we can actually process our data.

Veracity refers to the uncertainty of the quality of data. Data quality may vary, with perfect data for some dates and missing datasets for others. This affects our data pipeline and how much we can invest in our data platforms since, even today, one out of three business leaders doesn't completely trust the information they use to make business decisions.

Finally, velocity is probably the most important defining characteristic of big data (other than the obvious volume attribute), and it refers to the fact that big datasets not only have large volumes of data but also grow at an accelerated pace. This makes traditional storage using—for example—indexing a difficult task.

The big data landscape

Big data has evolved into a complex ecosystem affecting every sector of the economy. Going from hype to unrealistic expectations and back to reality, we now have big data systems implemented and deployed in most *Fortune 1000* companies that deliver real value.

If we segmented the companies that participate in the big data landscape by industry, we would probably come up with the following sections:

- Infrastructure
- Analytics

- Applications-enterprise

- Applications-industry

- Cross-infrastructure analytics

- Data sources and **application programming interfaces (APIs)**

- Data resources

- Open source

From an engineering point of view, we are probably more concerned about the underlying technologies than their applications in different industry sectors.

Depending on our business domain, we may have data coming in from different sources, such as transactional databases, **Internet of Things (IoT)** sensors, application server logs, other websites via a web service API, or just plain web page content extraction, as illustrated in the following diagram:

Standard ETL Flow

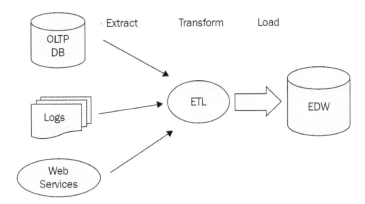

Figure 12.1: A sample ETL data flow

Message queuing systems

In most of the flows previously described, we have data being ETLed into an **enterprise data warehouse (EDW)**. To extract and transform this data, we need a message queuing system to deal with spikes in traffic, endpoints being temporarily unavailable, and other issues that may affect the availability and scalability of this part of the system.

Message queues also provide decoupling between producers and consumers of messages. This allows for better scalability by partitioning our messages into different topics/queues.

Finally, using message queues, we can have location-agnostic services that don't care where the message producers sit, which provides interoperability between different systems.

In the message queuing world, the most popular systems in production at the time of writing this book are RabbitMQ, ActiveMQ, and Kafka. We will provide a small overview of them before we dive into our use case to bring all of them together.

Apache ActiveMQ

Apache ActiveMQ is an open source message broker, written in Java, together with a full **Java Message Service** (**JMS**) client.

It is the most mature implementation out of the three that we examine here and has a long history of successful production deployments. Commercial support is offered by many companies, including Red Hat.

It is a fairly simple queuing system to set up and manage. It is based on the JMS client protocol and is the tool of choice for Java **Enterprise Edition** (**EE**) systems.

RabbitMQ

RabbitMQ, on the other hand, is written in Erlang and is based on **Advanced Message Queuing Protocol** (**AMQP**). AMQP is significantly more powerful and complicated than JMS, as it allows peer-to-peer messaging, request/reply, and **publish/subscribe** (**pub/sub**) models for one-to-one or one-to-many message consumption.

RabbitMQ has gained popularity in the past 5 years and is now the most searched-for queuing system.

RabbitMQ's architecture is outlined as follows:

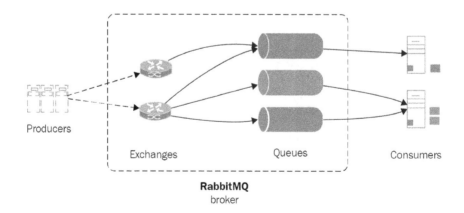

Figure 12.2: RabbitMQ architecture

Scaling in RabbitMQ systems is performed by creating a cluster of RabbitMQ servers. Clusters share data and state, which are replicated, but message queues are distinct per node. To achieve **high availability (HA)**, we can also replicate queues in different nodes.

Apache Kafka

Kafka, on the other hand, is a queuing system that was first developed by LinkedIn for its own internal purposes. It is written in Scala and is designed from the ground up for horizontal scalability and the best performance possible.

Focusing on performance is a key differentiator for Apache Kafka, but it means that in order to achieve performance, we need to sacrifice something. Messages in Kafka don't hold **unique identifiers (UIDs)** but are addressed by their offset in the log. Apache Kafka consumers are not tracked by the system; it is the responsibility of the application design to do so. Message ordering is implemented at the partition level, and it is the responsibility of the consumer to identify whether a message has been delivered already.

Semantics were introduced in version 0.11 and are part of the latest 1.0 release so that messages can now be both strictly ordered within a partition and always arrive exactly once for each consumer, as illustrated in the following diagram:

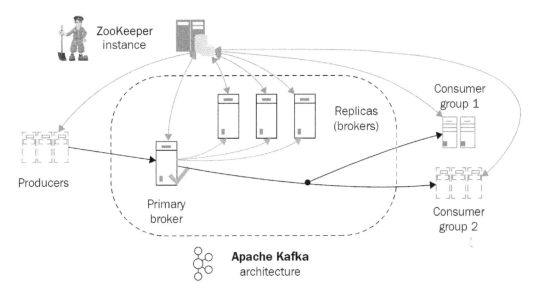

Figure 12.3: Apache Kafka architecture

The preceding diagram shows the overall Apache Kafka architecture. In the next section, we will learn about how to use MongoDB for data warehousing.

Data warehousing

Using a message queuing system is just the first step in our data pipeline design. At the other end of message queuing, we would typically have a data warehouse to process the vast amount of data that arrives. There are numerous options there, and it is not the main focus of this book to go over these or compare them. However, we will skim through two of the most widely used options from the **Apache Software Foundation** (**ASF**): Apache Hadoop and Apache Spark.

Apache Hadoop

The first, and probably still most widely used, framework for big data processing is Apache Hadoop. Its foundation is HDFS. Developed at Yahoo! in the 2000s, it originally served as an open source alternative to **Google File System** (**GFS**), a distributed filesystem that was serving Google's needs for distributed storage of its search index.

Hadoop also implemented a MapReduce alternative to Google's proprietary system, Hadoop MapReduce. Together with HDFS, they constitute a framework for distributed storage and computations. Written in Java, with bindings for most programming languages and many projects that provide abstracted and simple functionality, and sometimes based on **Structured Query Language** (**SQL**) querying, it is a system that can reliably be used to store and process TB, or even **petabytes** (**PB**), of data.

In later versions, Hadoop became more modularized by introducing **Yet Another Resource Negotiator** (**YARN**), which provides the abstraction for applications to be developed on top of Hadoop. This has enabled several applications to be deployed on top of Hadoop, such as **Storm**, **Tez**, **Open MPI**, **Giraph**, and, of course, **Apache Spark**, as we will see in the following sections.

Hadoop MapReduce is a batch-oriented system, meaning that it relies on processing data in batches, and is not designed for real-time use cases.

Apache Spark

Apache Spark is a cluster-computing framework from the University of California, Berkeley's AMPLab. Spark is not a substitute for the complete Hadoop ecosystem, but mostly for the MapReduce aspect of a Hadoop cluster. Whereas Hadoop MapReduce uses on-disk batch operations to process data, Spark uses both in-memory and on-disk operations. As expected, it is faster with datasets that fit in memory. This is why it is more useful for real-time streaming applications, but it can also be used with ease for datasets that don't fit in memory.

Apache Spark can run on top of HDFS using YARN or in standalone mode, as shown in the following diagram:

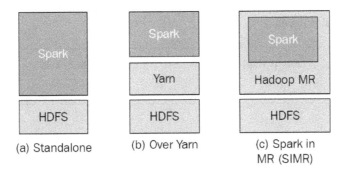

Figure 12.4: Apache Spark over HDFS

This means that in some cases (such as the one that we will use in our following use case), we can completely ditch Hadoop for Spark if our problem is really well defined and constrained within Spark's capabilities.

Spark can be up to 100 times faster than Hadoop MapReduce for in-memory operations. Spark offers user-friendly APIs for Scala (its native language), Java, Python, and Spark SQL (a variation of the *SQL-92* specification). Both Spark and MapReduce are resilient to failure. Spark uses **resilient distributed datasets (RDDs)** that are distributed across the whole cluster.

As we can see from the overall Spark architecture, as follows, we can have several different modules of Spark working together for different needs, from SQL querying to streaming and **machine learning (ML)** libraries.

Comparing Spark with Hadoop MapReduce

The Hadoop MapReduce framework is more commonly compared to Apache Spark, a newer technology that aims to solve problems in a similar problem space. Some of their most important attributes are summarized in the following table:

	Hadoop MapReduce	Apache Spark
Written in	Java	Scala
Programming model	MapReduce	RDD
Client bindings	Most high-level languages	Java, Scala, Python
Ease of use	Moderate, with high-level abstractions (Pig, Hive, and so on)	Good
Performance	High throughput in batch	High throughput in streaming and batch mode
Uses	Disk (**input/output (I/O)** bound)	Memory, degrading performance if a disk is needed
Typical node	Medium	Medium-large

Table 12.1: Apache Spark versus Hadoop MapReduce framework comparison

As we can see from the preceding comparison, there are pros and cons for both technologies. Spark arguably has better performance, especially in problems that use fewer nodes. On the other hand, Hadoop is a mature framework with excellent tooling on top of it to cover almost every use case.

MongoDB as a data warehouse

Apache Hadoop is often described as the 800-pound gorilla in the room of big data frameworks. Apache Spark, on the other hand, is more like a 200-pound cheetah for its speed, agility, and performance characteristics, which allow it to work well in a subset of the problems that Hadoop aims to solve.

MongoDB, on the other hand, can be described as the MySQL equivalent in the NoSQL world, because of its adoption and ease of use. MongoDB also offers an aggregation framework, MapReduce capabilities, and horizontal scaling using sharding, which is essentially data partitioning at the database level. So, naturally, some people wonder why we don't use MongoDB as our data warehouse to simplify our architecture.

This is a pretty compelling argument, and it may or may not be the case that it makes sense to use MongoDB as a data warehouse. The advantages of such a decision are set out here:

- Simpler architecture
- Less need for message queues, reducing latency in our system

The disadvantages are presented here:

- MongoDB's MapReduce framework is not a replacement for Hadoop's MapReduce. Even though they both follow the same philosophy, Hadoop can scale to accommodate larger workloads.
- Scaling MongoDB's document storage using sharding will lead to hitting a wall at some point. Whereas Yahoo! has reported using 42,000 servers in its largest Hadoop cluster, the largest MongoDB commercial deployments stand at 5 billion (*Craigslist*), compared to 600 nodes and PB of data for Baidu, the internet giant dominating, among others, the Chinese internet search market.

There is more than an **order of magnitude (OOM)** of difference in terms of scaling.

MongoDB is mainly designed around being a real-time querying database based on stored data on disk, whereas MapReduce is designed around using batches, and Spark is designed around using streams of data.

Big data use case with servers on-premises

Putting all of this into action, we will develop a fully working system using a data source, a Kafka message broker, an Apache Spark cluster on top of HDFS feeding a Hive table, and a MongoDB database. Our Kafka message broker will ingest data from an API, streaming market data for a **Monero (XMR)/ Bitcoin (BTC)** currency pair. This data will be passed on to an Apache Spark algorithm on HDFS to calculate the price for the next ticker timestamp, based on the following factors:

- The corpus of historical prices already stored on HDFS
- The streaming market data arriving from the API

This predicted price will then be stored in MongoDB using the MongoDB Connector for Hadoop. MongoDB will also receive data straight from the Kafka message broker, storing it in a special collection with the document expiration date set to 1 minute. This collection will hold the latest orders, with the goal of being used by our system to buy or sell, using the signal coming from the Spark ML system.

So, for example, if the price is currently 10 and we have a bid for 9.5, but we expect the price to go down at the next market tick, then the system would wait. If we expect the price to go up in the next market tick, then the system would increase the bid price to 10.01 to match the price in the next ticker.

Similarly, if the price is 10 and we bid for 10.5, but expect the price to go down, we would adjust our bid to 9.99 to make sure we don't overpay for it. But if the price is expected to go up, we would immediately buy to make a profit at the next market tick.

Schematically, our architecture looks like this:

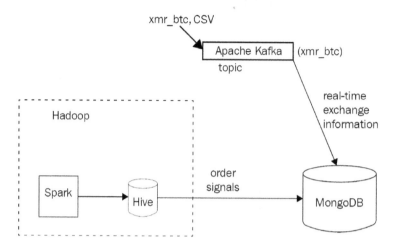

Figure 12.5: Sample use case Apache Kafka architecture

The API is simulated by posting **JavaScript Object Notation (JSON)** messages to a Kafka topic named `xmr_btc`. On the other end, we have a Kafka consumer importing real-time data to MongoDB.

We also have another Kafka consumer importing data to Hadoop to be picked up by our algorithms, which send recommendation data (signals) to a Hive table. Finally, we export data from the Hive table into MongoDB.

Setting up Kafka

The first step in setting up the environment for our big data use case is to establish a Kafka node. Kafka is essentially a **first-in, first-out (FIFO)** queue, so we will use the simplest single node (broker) setup. Kafka organizes data using topics, producers, consumers, and brokers.

The important Kafka terminologies are noted here:

- A **broker** is essentially a node
- A **producer** is a process that writes data to the message queue
- A **consumer** is a process that reads data from the message queue
- A **topic** is a specific queue that we write to and read data from

A Kafka topic is further subdivided into a number of partitions. We can split data from a particular topic into multiple brokers (nodes), both when we write to the topic and also when we read our data at the other end of the queue.

After installing Kafka on our local machine, or any cloud provider of our choice (there are excellent tutorials for **Elastic Compute Cloud** (**EC2**) to be found just a search away), we can create a topic using this single command:

```
$ kafka-topics --create --zookeeper localhost:2181
--replication-factor 1 --partitions 1 --topic xmr-btc
Created topic "xmr-btc".
```

This will create a new topic called xmr-btc.

Deleting the topic is similar to creating one, by using this command:

```
$ kafka-topics --delete --zookeeper localhost:2181 --topic
xmr-btc
```

We can then get a list of all topics by issuing the following command:

```
$ kafka-topics --list --zookeeper localhost:2181
xmr-btc
```

We can then create a command-line producer for our topic, just to test that we can send messages to the queue, like this:

```
$ kafka-console-producer --broker-list localhost:9092 --topic
xmr-btc
```

Data on every line will be sent as a string-encoded message to our topic, and we can end the process by sending a SIGINT signal (typically *Ctrl + C*).

Afterward, we can view messages that are waiting in our queue by spinning up a consumer, as follows:

```
$ kafka-console-consumer --zookeeper localhost:2181 --topic
xmr-btc --from-beginning
```

This consumer will read all messages in our xmr-btc topic, starting from the beginning of history. This is useful for our test purposes, but we will change this configuration in real-world applications.

> **Note**
>
> You will keep seeing zookeeper, in addition to kafka, mentioned in the commands. Apache ZooKeeper comes together with Apache Kafka and is a centralized service that is used internally by Kafka for maintaining configuration information, naming, providing distributed synchronization, and providing group services.

Now that we have set up our broker, we can use the code at `https://github.com/PacktPublishing/Mastering-MongoDB-6.x/tree/main/chapter_12` to start reading (consuming) and writing (producing) messages to the queue. For our purposes, we are using the `ruby-kafka` gem, developed by Zendesk.

For simplicity, we are using a single class to read from a file stored on disk and to write to our Kafka queue.

Our `produce` method will be used to write messages to Kafka, as follows:

```
def produce
options = { converters: :numeric, headers: true }
CSV.foreach('xmr_btc.csv', options) do |row|
json_line = JSON.generate(row.to_hash)
@kafka.deliver_message(json_line, topic: 'xmr-btc')
end
end
```

Our `consume` method will read messages from Kafka, as follows:

```
def consume
consumer = @kafka.consumer(group_id: 'xmr-consumers')
consumer.subscribe('xmr-btc', start_from_beginning: true)
trap('TERM') { consumer.stop }
consumer.each_message(automatically_mark_as_processed: false)
do |message|
puts message.value
if valid_json?(message.value)
MongoExchangeClient.new.insert(message.value)
consumer.mark_message_as_processed(message)
end
end
consumer.stop
end
```

> **Note**
>
> Notice that we are using the consumer group API feature (added in Kafka 0.9) to get multiple consumers to access a single topic by assigning each partition to a single consumer. In the event of a consumer failure, its partitions will be reallocated to the remaining members of the group.

The next step is to write these messages to MongoDB, as follows:

1. First, we create a collection so that our documents expire after 1 minute. Enter the following code in the `mongo` shell:

```
> use exchange_data
> db.xmr_btc.createIndex( { "createdAt": 1 }, {
expireAfterSeconds: 60 })
{
"createdCollectionAutomatically" : true,
"numIndexesBefore" : 1,
"numIndexesAfter" : 2,
"ok" : 1
}
```

This way, we create a new database called `exchange_data` with a new collection called `xmr_btc` that has auto-expiration after 1 minute. For MongoDB to auto-expire documents, we need to provide a field with a `datetime` value to compare its value against the current server time. In our case, this is the `createdAt` field.

2. For our use case, we will use the low-level MongoDB Ruby driver. The code for `MongoExchangeClient` is shown here:

```
class MongoExchangeClient
def initialize
@collection = Mongo::Client.new([ '127.0.0.1:27017' ],
database: :exchange_data).database[:xmr_btc]
end
def insert(document)
document = JSON.parse(document)
document['createdAt'] = Time.now
@collection.insert_one(document)
end
end
```

This client connects to our local database, sets the `createdAt` field for the **time-to-live** (TTL) document expiration, and saves the message to our collection.

With this set up, we can write messages to Kafka, read them at the other end of the queue, and write them into our MongoDB collection.

Setting up Hadoop

We can install Hadoop and use a single node for the use case in this chapter using the instructions from Apache Hadoop's website at `https://hadoop.apache.org/docs/stable/hadoop-project-dist/hadoop-common/SingleCluster.html`.

After following these steps, we can browse the HDFS files in our local machine at `http://localhost:50070/explorer.html#/`. Assuming that our signals data is written in HDFS under the `/user/<username>/signals` directory, we will use the MongoDB Connector for Hadoop to export and import it into MongoDB.

The MongoDB Connector for Hadoop is the officially supported library, allowing MongoDB data files or MongoDB backup files in **Binary JSON** (**BSON**) to be used as the source or destination for Hadoop MapReduce tasks.

This means that we can also easily export to, and import data from, MongoDB when we are using higher-level Hadoop ecosystem tools such as Pig (a procedural high-level language), Hive (a SQL-like high-level language), and Spark (a cluster-computing framework).

Steps for Hadoop setup

The different steps to set up Hadoop are set out here:

1. Download the **Java ARchive** (**JAR**) file from the Maven repository at `https://repo1.maven.org/maven2/org/mongodb/mongo-hadoop/mongo-hadoop-core/2.0.2/`.

2. Download `mongo-java-driver` from `https://oss.sonatype.org/content/repositories/releases/org/mongodb/mongodb-driver/3.5.0/`.

3. Create a directory (in our case, named `mongo_lib`) and copy these two JARs in there with the following command:

    ```
    export HADOOP_CLASSPATH=$HADOOP_CLASSPATH:<path_to_
    directory>/mongo_lib/
    ```

    Alternatively, we can copy these JARs under the `share/hadoop/common/` directory. As these JARs will need to be available in every node, for clustered deployment, it's easier to use Hadoop's `DistributedCache` to distribute the JARs to all nodes.

4. The next step is to install Hive from `https://hive.apache.org/downloads.html`. For this example, we used a MySQL server for Hive's metastore data. This can be a local MySQL server for development, but it is recommended that you use a remote server for production environments.

5. Once we have Hive set up, we just run the following command:

    ```
    > hive
    ```

6. Then, we add the three JARs (mongo-hadoop-core, mongo-hadoop-driver, and mongo-hadoop-hive) that we downloaded earlier, as follows:

```
hive> add jar /Users/dituser/code/hadoop-2.8.1/mongo-
hadoop-core-2.0.2.jar;
Added [/Users/dituser/code/hadoop-2.8.1/mongo-hadoop-
core-2.0.2.jar] to class path
Added resources: [/Users/dituser/code/hadoop-2.8.1/mongo-
hadoop-core-2.0.2.jar]
hive> add jar /Users/dituser/code/hadoop-2.8.1/mongodb-
driver-3.5.0.jar;
Added [/Users/dituser/code/hadoop-2.8.1/mongodb-driver-
3.5.0.jar] to class path
Added resources: [/Users/dituser/code/hadoop-2.8.1/
mongodb-driver-3.5.0.jar]
hive> add jar /Users/dituser/code/hadoop-2.8.1/mongo-
hadoop-hive-2.0.2.jar;
Added [/Users/dituser/code/hadoop-2.8.1/mongo-hadoop-
hive-2.0.2.jar] to class path
Added resources: [/Users/dituser/code/hadoop-2.8.1/mongo-
hadoop-hive-2.0.2.jar]
hive>
```

And then, we assume our data is in the table exchanges shown here:

customerid	Int
pair	String
time	TIMESTAMP
recommendation	Int

Table 12.2: Data types

> **Note**
>
> We can also use Gradle or Maven to download the JARs in our local project. If we only need MapReduce, then we just download the mongo-hadoop-core JAR. For Pig, Hive, Streaming, and so on, we must download the appropriate JARs from https://repo1.maven.org/maven2/org/mongodb/mongo-hadoop/.
>
> Some useful Hive commands include the following: show databases; and create table exchanges(customerid int, pair String, time TIMESTAMP, recommendation int);.

7. Now that we are all set, we can create a MongoDB collection backed by our local Hive data, as follows:

```
hive> create external table exchanges_mongo (objec-
tid STRING, customerid INT,pair STRING,time STRING,
recommendation INT) STORED BY 'com.mongodb.hadoop.
hive.MongoStorageHandler' WITH SERDEPROPERTIES('mongo.
columns.mapping'='{"objectid":"_id","customerid":"cus-
tomerid","pair":"pair","time":"Timestamp", "recommenda-
tion":"recommendation"}') tblproperties('mongo.uri'='mon-
godb://localhost:27017/exchange_data.xmr_btc');
```

8. Finally, we can copy all data from the exchanges Hive table into MongoDB, as follows:

```
hive> Insert into table exchanges_mongo select * from
exchanges;
```

This way, we have established a pipeline between Hadoop and MongoDB using Hive, without any external server.

Using a Hadoop-to-MongoDB pipeline

An alternative to using the MongoDB Connector for Hadoop is to use the programming language of our choice to export data from Hadoop, and then write into MongoDB using the low-level driver or an **object document mapper** (**ODM**), as described in previous chapters.

For example, in Ruby, there are a few options, as follows:

- **WebHDFS** on GitHub, which uses the WebHDFS or the **HttpFS** Hadoop API to fetch data from HDFS

- System calls, using the Hadoop command-line tool and Ruby's system() call

Whereas in Python, we can use the following:

- **HdfsCLI**, which uses the WebHDFS or the HttpFS Hadoop API

- **libhdfs**, which uses a **Java Native Interface** (**JNI**)-based native C wrapped around the HDFS Java client

All of these options require an intermediate server between our Hadoop infrastructure and our MongoDB server, but, on the other hand, allow for more flexibility in the ETL process of exporting/importing data.

Setting up a Spark connection to MongoDB

MongoDB also offers a tool to directly query Spark clusters and export data to MongoDB. Spark is a cluster computing framework that typically runs as a YARN module in Hadoop, but can also run independently on top of other filesystems.

The MongoDB Spark Connector can read and write to MongoDB collections from Spark using Java, Scala, Python, and R. It can also use aggregation and run SQL queries on MongoDB data after creating a temporary view for the dataset backed by Spark.

Using Scala, we can also use Spark Streaming, the Spark framework for data-streaming applications built on top of Apache Spark.

In the next section, we will discuss how to use data warehousing in the cloud, using the MongoDB Atlas Data Lake service.

MongoDB Atlas Data Lake

A data lake is a centralized repository that can be used to store, query, and transform heterogeneous structured and unstructured data. As the name implies, it acts as a lake of data.

MongoDB Data Lake (`https://www.mongodb.com/atlas/data-lake`) is a service offered by MongoDB that can help us process datasets across multiple MongoDB Atlas clusters and **Amazon Web Services (AWS) Simple Storage Service (S3)** buckets.

A data lake can query data in multiple formats such as JSON, BSON, **comma-separated values (CSV)**, **tab-separated values (TSV)**, Avro, **Optimized Row Columnar (ORC)**, and Parquet. We can query the datasets using any driver, **graphical user interface (GUI)** tools such as MongoDB Compass, or the MongoDB shell using the standard **MongoDB Query Language (MQL)**.

We can use x509 or SCRAM-SHA authentication methods. MongoDB Data Lake does not support **Lightweight Directory Access Protocol (LDAP)**.

A data lake can be used for a variety of big data use cases, including but not limited to the following:

- Querying data across multiple MongoDB Atlas clusters to generate business insights that were previously uncovered.

- Converting MongoDB data into one of the supported formats—for example, Parquet or Avro or CSV. This could be part of an ETL process to extract data from a MongoDB Atlas cluster into another system for reporting, ML, or other reasons.

- Ingest data into our MongoDB Atlas cluster from an S3 bucket. This could again be part of an ETL process.

- Process and store aggregations performed against MongoDB Atlas or S3-hosted data.

We need to grant read-only or read-write access to the S3 buckets that we want to use with MongoDB Data Lake. Atlas will use existing **role-based access control** (**RBAC**) to access the MongoDB clusters.

MongoDB Atlas Data Lake is the recommended approach for cloud-based big data systems, offering features and support by MongoDB.

Summary

In this chapter, we learned about the big data landscape and how MongoDB compares with, and fares against, message-queuing systems and data warehousing technologies. Using a big data use case, we learned how to integrate MongoDB with Kafka and Hadoop from a practical perspective. Finally, we learned about data lakes and how we can use them to store and process data from different data sources such as MongoDB Atlas and Amazon S3 buckets.

In the next chapter, we will turn to replication and cluster operations, and discuss replica sets, the internals of elections, and the setup and administration of our MongoDB cluster.

Further reading

You can refer to the following references for further information:

- Internet adoption estimates: `https://www.cisco.com/c/en/us/solutions/collateral/service-provider/visual-networking-index-vni/vni-hyperconnectivity-wp.html`

- Big data companies landscape (2017): `http://mattturck.com/wp-content/uploads/2017/05/Matt-Turck-FirstMark-2017-Big-Data-Landscape.png`

- Message queuing explained: `https://www.cloudamqp.com/blog/2014-12-03-what-is-message-queuing.html`

- JMS and AMQP message queuing systems: `https://www.linkedin.com/pulse/jms-vs-amqp-eran-shaham`

- Apache Kafka and RabbitMQ: `https://www.cloudamqp.com/blog/2017-01-09-apachekafka-vs-rabbitmq.html`

- MapReduce Google paper: `https://static.googleusercontent.com/media/research.google.com/en//archive/mapreduce-osdi04.pdf`

- Apache Hadoop architecture: `https://en.wikipedia.org/wiki/Apache_Hadoop#Architecture`

- YARN and MapReduce 2: `https://www.slideshare.net/cloudera/introduction-to-yarn-and-mapreduce-2?next_slideshow=1`

- MongoDB at Craigslist (2011): `https://www.mongodb.com/blog/post/mongodb-live-at-craigslist`

- MongoDB at Baidu (2016): `https://www.mongodb.com/blog/post/mongodb-at-baidu-powering-100-apps-across-600-nodes-at-pb-scale`

- Hadoop and Spark: `http://www.datamation.com/data-center/hadoop-vs.-spark-the-new-age-of-big-data.html`

- *MongoDB as a Data Warehouse*: `https://www.mongodb.com/mongodb-data-warehouse-time-series-and-device-history-data-medtronic-transcript`

- Hadoop and Apache Spark:

  `https://www.infoworld.com/article/3014440/big-data/five-things-you-need-to-know-about-hadoop-v-apache-spark.html`

  `https://iamsoftwareengineer.wordpress.com/2015/12/15/hadoop-vs-spark/?iframe=true&theme_preview=true`

- Apache Kafka: `https://www.infoq.com/articles/apache-kafka`

- Apache Kafka exactly-once message semantics: `https://medium.com/@jaykreps/exactly-once-support-in-apache-kafka-55e1fdd0a35f`

- Apache Kafka and RabbitMQ: `https://www.slideshare.net/sbaltagi/apache-kafka-vs-rabbitmq-fit-for-purpose-decision-tree`

- Kafka Ruby library: `https://github.com/zendesk/ruby-kafka`

- Apache Hadoop on Mac OS X tutorial: `http://zhongyaonan.com/hadoop-tutorial/setting-up-hadoop-2-6-on-mac-osx-yosemite.html`

- API and **command-line interface (CLI)** for HDFS: `https://github.com/mtth/hdfs`

- Python HDFS library: `http://wesmckinney.com/blog/python-hdfs-interfaces/`

- *Kafka in a Nutshell*: `https://sookocheff.com/post/kafka/kafka-in-a-nutshell/`

- Apache Spark and Python using MLlib: `https://www.codementor.io/jadianes/spark-mllib-logistic-regression-du107neto`

- Apache Spark and Ruby: `http://ondra-m.github.io/ruby-spark/`

- Apache Spark introduction: `https://www.infoq.com/articles/apache-spark-introduction`

- Apache Spark Stanford paper: `https://cs.stanford.edu/~matei/papers/2010/hotcloud_spark.pdf`

Part 4 – Scaling and High Availability

The last part of the book will cover replication and sharding, which are hugely important for production deployments of sufficient data size. In the last chapter, we will provide several recommendations and pointers for cluster administration and maintenance, with a focus on high availability and fault tolerance.

This part contains the following chapters:

- Chapter 13, *Mastering Replication*
- Chapter 14, *Mastering Sharding*
- Chapter 15, *Fault Tolerance and High Availability*

13

Mastering Replication

Replication has been one of the most useful features of MongoDB since the very early days. In general, replication refers to the process of synchronizing data across different servers. The benefits of replication include protection from data loss and high availability of data. Replication also provides disaster recovery, avoidance of downtime for maintenance, and scaling reads since we can read from the primary and any of the secondary servers. In this chapter, we will cover the following topics:

- Replication

- An architectural overview

- How do elections work?

- What is the use case for a replica set?

- Setting up a replica set

- Connecting to a replica set

- Replica set administration

- Cloud options for a replica set

- Replica set limitations

By the end of this chapter, you will have mastered the underlying theory and concepts as well as the practical implementation of replication using MongoDB.

Technical requirements

To follow along with the code in this chapter, you need to install MongoDB locally or connect to a MongoDB Atlas database. You can download the MongoDB community edition from mongodb. com or use the fully managed DBaaS MongoDB Atlas offering, which provides a free tier as well as seamless upgrades to the latest version.

Replication

There are different approaches to replication. The approach that MongoDB takes is logical replication with a primary-secondary, which we will explain in more detail later in this chapter.

Logical or physical replication

With replication, we synchronize data across multiple servers, providing data availability and redundancy. Even if we lose a server due to a hardware or software failure, by using replication, we will have multiple copies that we can use to restore our data. Another advantage of replication is that we can use one of the servers as a dedicated reporting, or backup, server.

In logical replication, we have our primary server performing operations; the secondary server tails a queue of operations from the primary and applies the same operations in the same order. Using MongoDB as an example, the **operations log** (**oplog**) keeps track of operations that have happened on the primary server and applies them in the exact same order on the secondary server.

Logical replication is useful for a wide array of applications, such as information sharing, data analysis, and **Online Analytical Processing** (**OLAP**) reporting.

In physical replication, data gets copied on the physical level, at a lower level than database operations. This means that we are not applying the operations but copying the bytes that were affected by these operations. It also means that we can gain better efficiency since we are using low-level structures to transfer data. We can also ensure that the state of the database is exactly the same, since they are identical, byte for byte.

What is typically missing from physical replication is knowledge about the database structure, which means that it is harder (if not impossible) to copy some collections from a database and ignore others.

Physical replication is typically suited for more rare circumstances, such as disaster recovery, wherein a full and exact copy of everything (including data, indexes, the internal state of the database in a journal, and redoing/undoing logs) is of crucial importance to bringing the application back to the exact state it was in.

Different high availability types

In high availability, there are several configurations that we can use. Our primary server is called the **hot server** as it can process each and every request coming in. A secondary server can be in any of the following states:

- Cold
- Warm
- Hot

A **secondary cold server** is a server that is there just in case the primary server goes offline, without any expectation of it holding the data and state that the primary server had.

A **secondary warm server** receives periodic updates of data from the primary server, but typically, it is not entirely up to date with the primary server. It can be used for some non-real-time analytics reporting to offload the main server, but typically, it will not be able to pick up the transactional load of the primary server if it goes down.

A **secondary hot server** always keeps an up-to-date copy of the data and state from the primary server. It usually waits in a hot standby state, ready to take over when the primary server goes down.

MongoDB has both the hot and warm server types of functionality, as we will explore in the following sections.

> **Note**
>
> Most database systems employ a similar notion of primary/secondary servers, so conceptually, everything from MongoDB gets applied there, too.

An architectural overview

MongoDB's replication is illustrated in the following diagram:

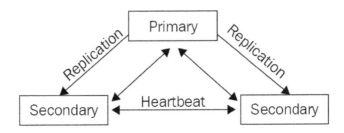

Figure 13.1 – MongoDB replication conceptual architecture

The primary server is the only one that can take writes at any time. The secondary servers are in a hot standby state, ready to take over if the primary server fails. Once the primary server fails, an election takes place regarding which secondary server will become primary.

We can also have at most one **arbiter node**. An arbiter node does not hold any data, and its sole purpose is to participate in the election process.

We must always have an odd number of nodes (including the arbiter). Three, five, and seven are all fine so that in the event of the primary (or more servers) failing, we have a majority of votes in the election process.

When the other members of a replica set don't hear from the primary for more than 10 seconds (configurable), an eligible secondary will start the election process to vote for a new primary. The first secondary to hold the election and win the majority will become the new primary. All remaining servers will now replicate from the new primary server, keeping their roles as secondaries but syncing up from the new primary.

Starting with MongoDB 3.6, client drivers can retry write operations a **single time** if they detect that the primary is down. A replica set can have up to 50 members, but only up to 7 of them can vote in the election process.

The setup for our replica set after the new election will be as follows:

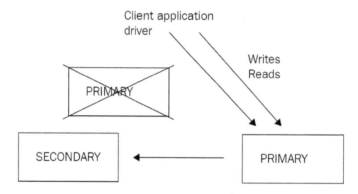

Figure 13.2 – MongoDB replica set election process

In the following section, we will discuss how elections work.

How do elections work?

All of the servers in a replica set maintain regular communication with every other member via a heartbeat. The heartbeat is a small packet that's regularly sent to verify that all members are operating normally.

Secondary members also communicate with the primary to get the latest updates from the oplog and apply them to their own data.

> **Note**
>
> The information here refers to the latest replication election protocol, version 1, which is the latest as of MongoDB v6.0.

Schematically, we can see how this works.

When the primary member goes down, all of the secondaries will miss a heartbeat or more. They will be waiting up until the `settings.electionTimeoutMillis` time passes (the default is 10 seconds), and then the secondaries will start one or more rounds of elections to find the new primary.

For a server to be elected as primary from the secondaries, it must have the following two properties:

- Belong in a group of voters that have *50% + 1* of the votes
- Be the most up-to-date secondary in this group

In a simple example of three servers with one vote each, once we lose the primary, the other two servers will each have one vote (so, in total, two-thirds), and as such, the one with the most up-to-date oplog will be elected as primary.

Now, consider a more complex setup, as follows:

- Seven servers (one primary, six secondaries)
- One vote each

We lose the primary server, and the six remaining servers have network connectivity issues, resulting in a network partition as follows:

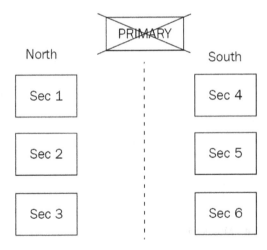

Figure 13.3 – MongoDB replica set network partitioning

These partitions can be described as follows:

- Partition **North**: Three servers (one vote each)
- Partition **South**: Three servers (one vote each)

Neither partition has any knowledge of what happened to the rest of the servers. Now, when they hold elections, no partition can establish a majority as they have three out of seven votes. No primary will get elected from either partition. This problem can be overcome by having, for example, one server with three votes.

Now, our overall cluster setup looks as follows:

- **Server #1**: One vote
- **Server #2**: One vote
- **Server #3**: One vote
- **Server #4**: One vote
- **Server #5**: One vote
- **Server #6**: One vote
- **Server #7**: Three votes

After losing Server #1, our partitions now look as follows:

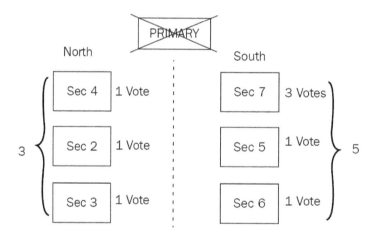

Figure 13.4 – MongoDB replica set network partitioning voting

Partition **North** is as follows:

- **Server #2**: One vote
- **Server #3**: One vote
- **Server #4**: One vote

Partition **South** is as follows:

- **Server #5**: One vote

- **Server #6**: One vote

- **Server #7**: Three votes

Partition **South** has three servers, with a total of five out of nine votes. The secondary among servers #5, #6, and #7 that is most up to date (according to its oplog entries) will be elected as the primary.

After learning the fundamental concepts behind replication architecture, we will discuss when we should use a replica set in the following section.

What is the use case for a replica set?

MongoDB offers most of the advantages of using a replica set, some of which are listed as follows:

- Protection from data loss

- High availability of data

- Disaster recovery

- Avoidance of downtime for maintenance

- Scaling reads since we can read from multiple servers

- Helping to design for geographically dispersed services

- Data privacy

The most notable item that's missing from the list is scaling writes. This is because, in MongoDB, we can only have one primary, and only this primary can take writes from our application server.

When we want to scale write performance, we typically design and implement sharding, which will be the topic of the following chapter. Two interesting properties of the way that MongoDB replication is implemented are geographically dispersed services and data privacy.

It is not uncommon for our application servers to be located in multiple data centers across the globe. Using replication, we can have a secondary server as close to the application server as possible. What this means is that our reads will be fast, as if they were local, and we will get a latency performance penalty just for our writes. This requires some planning at the application level, of course, so that we can maintain two different pools of connections to our database, which can be easily done by either using the official MongoDB drivers or using higher-level ODMs.

The second interesting property of MongoDB's replication design is implementing data privacy. When we have servers geographically dispersed across different data centers, we can enable replication per database. By keeping a database out of the replication process, we can make sure that our data stays confined to the data center that we need. We can also set up different replication schemas per database in the same MongoDB server so that we have multiple replication strategies according to our data privacy needs, excluding some servers from our replica sets if they are not allowed by our data privacy regulations.

Setting up a replica set

In this section, we will go over the most common deployment procedures to set up a replica set. These involve either converting a standalone server into a replica set or setting up a replica set from scratch.

Converting a standalone server into a replica set

To convert a standalone server into a replica set, we first need to cleanly shut down the mongo server as follows:

```
> use admin
> db.shutdownServer()
```

Then, we start the server with the --replSet configuration option via the command line as follows (or by using a configuration file, as we will explain in the following section):

1. First, we connect (via the mongo shell) to the new replica set-enabled instance as follows:

    ```
    > rs.initiate()
    ```

2. Now, we have the first server of our replica set. We can add the other servers (which must have also been started with --replSet) by using the mongo shell as follows:

    ```
    > rs.add("<hostname><:port>")
    ```

> **Note**
>
> Double-check the replica set configuration by using rs.conf(). Verify the replica set status by using rs.status().

Creating a replica set

Starting a MongoDB server as a part of a replica set is as easy as setting it in the configuration via the command line as follows:

```
> mongod --replSet ''xmr_cluster''
```

This is fine for development purposes. For production environments, it's recommended that we use a configuration file instead, as follows:

```
> mongod --config <path-to-config>
```

Here, `<path-to-config>` can be as follows:

```
/etc/mongod.conf
```

This configuration file has to be in a YAML format.

> **Note**
> YAML does not support tabs. Convert tabs to spaces by using your editor of choice.

A simple configuration file sample is as follows:

```
systemLog:
destination: file
path: "/var/log/mongodb/mongod.log"
logAppend: true
storage:
journal:
enabled: true
processManagement:
fork: true
net:
bindIp: 127.0.0.1
port: 27017
replication:
oplogSizeMB: <int>
replSetName: <string>
```

Root-level options define the sections that leaf-level options apply to by nesting. Regarding replication, the mandatory options are `oplogSizeMB` (the oplog size for the member, in MB) and `replSetName` (the replica set name, such as `xmr_cluster`).

We can also set the following on the same level as `replSetName`:

```
secondaryIndexPrefetch: <string>
```

This is only available for the MMAPv1 storage engine, and it refers to the indexes on secondaries that will get loaded into memory before applying operations from the oplog.

It defaults to all, and the available options are none and _id_only, in order to load no indexes into memory and only load the default index that was created on _id fields as follows:

```
enableMajorityReadConcern: <boolean>
```

This is the configuration setting for enabling the read preference of majority for this member.

After we have started all of the replica set processes on different nodes, we log in to one of the nodes using mongo from the command line with the appropriate host:port. Then, we need to initiate the cluster from one member.

We can use configuration files as follows:

```
> rs.initiate()
```

We can also pass in the configurations as a document parameter as follows:

```
> rs.initiate( {
 _id : ''xmr_cluster'',
 members: [ { _id : 0, host : ''host:port'' } ]
})
```

> **Note**
>
> We can verify that the cluster was initiated by using rs.conf() in the shell.

Following that, we add each other member to our replica set by using host:port, which we defined in our networking setup, as follows:

```
> rs.add(''host2:port2'')
> rs.add(''host3:port3'')
```

> **Note**
>
> The minimum number of servers that we must use for a high-availability replica set is 3. We could replace one of the servers with an arbiter, but this is not recommended. Once we have added all of the servers and have waited a bit, we can check the status of our cluster by using rs.status(). By default, the oplog will be 5% of the free disk space. If we want to define it when we create our replica set, we can do so by passing the --oplogSizeMB command-line parameteror replication.oplogSizeMB in our configuration file. An oplog size cannot be more than 50 GB.

Read preference

By default, all writes and reads go/come from the primary server. Secondary servers replicate data but are not used for querying.

In some cases, it may be beneficial to change this and start to take reads from secondaries.

The MongoDB official drivers support the following five levels of read preference:

Read Preference Mode	Description
primary	This is the default mode where reads come from the primary server of the replica set.
primaryPreferred	With this mode, applications will read from the primary unless it is unavailable, in which case reads will come from secondary members.
secondary	Reads come exclusively from secondary servers.
secondaryPreferred	With this mode, applications will read from secondary members unless they are unavailable, in which case reads will come from the primary member.
nearest	Applications will read from the member of the replica set that is nearest in terms of network latency, not taking into account the member's type.

Table 13.1 – Read preference levels

> **Note**
>
> When we use the multi-document transactions and we include at least one read operation in the transaction, then we need to use the **primary** read preference and route all operations to the same server.

Using any read preference other than primary can be beneficial for asynchronous operations that are not extremely time-sensitive. For example, reporting servers can take reads from secondaries instead of the primary, as we may be fine with a small delay in our aggregation data, with the benefit of incurring more read load on our primary server.

Geographically distributed applications will also benefit from reading from secondaries, as these will have significantly lower latency. Although it's probably counter-intuitive, just changing the read preference from primary to secondary will not significantly increase the total read capacity of our cluster. This is because all of the members of our cluster are taking the same write load from clients' writes and replication for the primary and secondaries, respectively.

More importantly, however, reading from a secondary may return stale data, which has to be dealt with at the application level. Reading from different secondaries that may have variable replication lag (compared to our primary writes) may result in reading documents out of their insertion order (**non-monotonic reads**).

With all of the preceding caveats, it is still a good idea to test reading from secondaries if our application design supports it. An additional configuration option that can help us to avoid reading stale data is maxStalenessSeconds.

Based on a coarse estimation from each secondary as to how far behind the primary it is, we can set this to a value of 90 (seconds) or more to avoid reading stale data. Given that secondaries know how far behind they are from the primary (but don't accurately or aggressively estimate it), this should be treated as an approximation rather than something we base our design on.

Write concern

By default, the write operations in MongoDB replica sets will be acknowledged once the write has been acknowledged by the majority of servers in the replica set, that is {w: "majority"}. If we want to change this behavior, we can do so in two different ways as follows:

- We can request a different write concern per operation, in cases where we want to make sure that a write has propagated to multiple members of our replica set before marking it as complete, as follows:

```
> db.mongo_books.insert(
  { name: "Mastering MongoDB", isbn: "1001" },
  { writeConcern: { w: 2, wtimeout: 5000 } }
)
```

In the preceding example, we are waiting for the write to be confirmed by two servers (the primary, plus any one of the secondaries). We are also setting a timeout of 5000 milliseconds to avoid our write from blocking in cases where the network is slow or we just don't have enough servers to acknowledge the request.

- We can also change the default write concern across the entire replica set, as follows:

```
> cfg = rs.conf()
> cfg.settings.getLastErrorDefaults = { w: "majority",
wtimeout: 5000 }
> rs.reconfig(cfg)
```

Here, we set the write concern to majority with a timeout of 5 seconds. The majority write concern makes sure that our writes will propagate to at least $n/2+1$ servers, where n is the number of replica set members.

> **Note**
>
> The `majority` write concern is useful if we have a read preference of `majority` as well, as it ensures that every write with `w: "majority"` will also be visible with the same read preference. If we set `w>1`, it's useful to also set `wtimeout: <milliseconds>` with it. `wtimeout` will return from our write operation once the timeout has been reached, thus not blocking our client for an indefinite period of time. It's recommended to set `j: true` as well. `j: true` will wait for our write operation to be written to the journal before acknowledging it. `w>1`, along with `j: true`, will wait for the number of servers that we have specified to write to the journal before the acknowledgment.

Custom write concerns

We can also identify our replica set members with different tags (that is, `reporting`, east coast servers, and HQ servers) and specify a custom write concern per operation, as follows:

1. Use the usual procedure to connect to the primary via the mongo shell as follows:

    ```
    > conf = rs.conf()
    > conf.members[0].tags = { "location": "UK", "use":
    "production", "location_uk":"true" }
    > conf.members[1].tags = { "location": "UK", "use":
    "reporting", "location_uk":"true" }
    > conf.members[2].tags = { "location": "Ireland", "use":
    "production" }
    ```

2. We can now set a custom write concern as follows:

    ```
    > conf.settings = { getLastErrorModes: { UKWrites : {
    "location_uk": 2} } }
    ```

3. After applying the preceding code, we use the `reconfig` command as follows:

    ```
    > rs.reconfig(conf)
    ```

4. We can now start by setting `writeConcern` in our writes as follows:

    ```
    > db.mongo_books.insert({<our insert object>}, {
    writeConcern: { w: "UKWrites" } })
    ```

This means that our write will only be acknowledged if the `UKWrites` write concern is satisfied, which, in turn, will be satisfied by at least two servers with the `location_uk` tag verifying it. Since we only have two servers located in the UK, we can make sure that with this custom write concern, we have written our data to all of our UK-based servers.

Priority settings for replica set members

MongoDB allows us to set different priority levels for each member. This allows for some interesting applications and topologies to be implemented.

To change the priority after we have set up our cluster, we have to connect to our primary using the mongo shell and get the configuration object (in this case, `cfg`) as follows:

```
> cfg = rs.conf()
```

Then, we can change the `members` subdocument's `priority` attribute to the value of our choice, as follows:

```
> cfg.members[0].priority = 0.778
> cfg.members[1].priority = 999.9999
```

> **Note**
>
> The default `priority` value is 1 for every member. The priority can be set from 0 (never become a primary) to 1000, in floating-point precision.

Higher-priority members will be the first to call an election when the primary steps down, and they are also the most likely to win the election.

> **Note**
>
> Custom priorities should be configured with consideration of the different network partitions. Setting priorities the wrong way may lead to elections not being able to elect a primary, thus stopping all writes to our MongoDB replica set.

If we want to prevent a secondary from becoming a primary, we can set its priority to 0, as we will explain in the following section.

Zero priority replica set members

In some cases (for example, if we have multiple data centers), we will want some of the members to never be able to become a primary server.

In a scenario with multiple data center replications, we may have our primary data center with one primary and one secondary based in the UK, and a secondary server located in Russia. In this case, we don't want our Russia-based server to become primary, as it would incur latency on our application servers based in the UK. In this case, we will set up our Russia-based server with `priority` as 0.

Replica set members with `priority` as 0 also can't trigger elections. In all other aspects, they are identical to every other member in the replica set. To change the `priority` value of a replica set member, we must first get the current replica set configuration by connecting (via the mongo shell) to the primary server as follows:

```
> cfg = rs.conf()
```

This will provide the config document that contains the configuration for every member in our replica set. In the `members` subdocument, we can find the `priority` attribute, which we have to set to 0, as follows:

```
> cfg.members[2].priority = 0
```

Finally, we need to reconfigure the replica set with the updated configuration:

```
rs.reconfig(cfg)
```

> **Note**
>
> Make sure that you have the same version of MongoDB running in every node, otherwise, there may be unexpected behavior. Avoid reconfiguring the replica set cluster during high-volume periods. Reconfiguring a replica set may force an election for a new primary, which will close all active connections and may lead to a downtime of 10-30 seconds. Try to identify the lowest traffic time window to run maintenance operations such as reconfiguration, and always have a recovery plan in case something goes wrong.

Hidden replica set members

Hidden replica set members are used for special tasks. They are invisible to clients, will not show up in the `db.isPrimary()` mongo shell command and similar administrative commands, and, for all purposes, will not be taken into account by clients (that is, read preference options).

They can vote for elections but will never become a primary server. A hidden replica set member will only sync up to the primary server and doesn't take reads from the clients. As such, it has the same write load as the primary server (for replication purposes), but no read load on its own.

Due to the aforementioned characteristics, reporting is the most common application of a hidden member. We can connect directly to this member and use it as the data source of truth for OLAP.

To set up a hidden replica set member, we follow a similar procedure of setting `priority` to 0. After we have connected to our primary via the mongo shell, we get the configuration object, identify the member in the members subdocument that corresponds to the member we want to set as `hidden`, and subsequently set `priority` to 0 and its `hidden` attribute to `true`. Finally, we have to apply the new configuration by calling `rs.reconfig(config_object)` with `config_object,` which we used as a parameter, as follows:

```
> cfg = rs.conf()
> cfg.members[0].priority = 0
> cfg.members[0].hidden = true
> rs.reconfig(cfg)
```

A `hidden` replica set member can also be used for backup purposes. However, as you will see in the following section, we may want to use other options, either at the physical level or to replicate data at the logical level. In those cases, consider using a delayed replica set instead.

Delayed replica set members

In many cases, we will want to have a node that holds a copy of our data at an earlier point in time. This helps to recover from a big subset of human errors, such as accidentally dropping a collection, or an upgrade going horrendously wrong.

A delayed replica set member must have `priority = 0` and `hidden = true`. A delayed replica set member can vote for elections but will never be visible to clients (`hidden = true`) and will never become a primary (`priority = 0`).

An example is as follows:

```
> cfg = rs.conf()
> cfg.members[0].priority = 0
> cfg.members[0].hidden = true
> cfg.members[0].secondaryDelaySecs = 7200
> rs.reconfig(cfg)
```

This will set `members[0]` to a delay of 2 hours. Two important factors for deciding on the delta time period between the primary and delayed secondary server are as follows:

- Enough oplog size in the primary
- Enough time for the maintenance to finish before the delayed member starts picking up data

The current version of MongoDB also supports growing the oplog size over the configured size limit to avoid getting out of sync with the primary. That being said, we should still plan for the right oplog size instead of relying on MongoDB.

The following table shows the delay of the replica set in hours:

Maintenance window, in hours	Delay	Oplog size on primary, in hours
0.5	[0.5,5]	5

Table 13.2 – Replica set delayed, member

Production considerations

Deploy each mongod instance on a separate physical host. If you are using VMs, make sure that they map to different underlying physical hosts. Use the bind_ip option to make sure that your server maps to a specific network interface and port address.

Use firewalls to block access to any other port and/or only allow access between application servers and MongoDB servers. Even better, set up a VPN so that your servers communicate with each other in a secure, encrypted fashion.

We need to ensure that we have enabled authentication in our MongoDB cluster before binding to any IP address other than localhost.

In the following section, we will learn how to connect and perform administration tasks with the replica set configuration.

Connecting to a replica set

Connecting to a replica set is not fundamentally different from connecting to a single server. In this section, we will show some examples that use the official mongo-ruby-driver. We will use the following steps for the replica set:

1. First, we need to set our host and options objects as follows:

    ```
    client_host = ['hostname:port']
    client_options = {
    database: 'signals',
    replica_set: 'xmr_btc'
    }
    ```

    In the preceding example, we are getting ready to connect to hostname:port in the database signals in replica_set xmr_btc.

2. Calling the initializer on `Mongo::Client` will now return a `client` object that contains a connection to our replica set and database, as follows:

```
client = Mongo::Client.new(client_host, client_options)
```

The `client` object has the same options it has when connecting to a single server.

> **Note**
>
> MongoDB uses auto-discovery after connecting to our `client_host` to identify the other members of our replica set, regardless of whether they are the primary or secondaries. The `client` object should be used as a singleton, created once, and reused across our code base.

3. Having a singleton `client` object is a rule that can be overridden in some cases. We should create different `client` objects if we have different classes of connections to our replica set.

    An example would be having a `client` object for most operations, and then another `client` object for operations that are fine with only reading from secondaries, as follows:

```
client_reporting = client.with(:read => { :mode =>
:secondary })
```

4. This Ruby MongoDB `client` command will return a copy of the `MongoDB:Client` object with a read preference secondary that can be used, for example, for reporting purposes.

Some of the most useful options that we can use in our `client_options` initialization object are as follows:

Option	Description	Type	Default
replica_set	As used in our example: the replica set name.	String	None
Write	The write concern options as a hash object; the available options are w, wtimeout, j, and fsync. That is, to specify writes to two servers, with journaling, flushing to disk (fsync) true, and a timeout of 1 second as follows: `{ write: { w: 2, j: true, wtimeout: 1000, fsync: true } }`	Hash	`{ w: 1 }`

read	The read preference mode as a hash. Available options are mode and tag_sets.
	
	That is, to limit reads from secondary servers that have the UKWrites tag as follows:
	
	`{ read:`
	`{ mode: :secondary,`
	`  tag_sets: [ "UKWrites"`
	`]`
	
	`}`
	
	`}`
		Hash	`{ mode: primary }`

user	The name of the user to authenticate with.	String	None
password	The password of the user to authenticate with.	String	None
connect	Using :direct, we can force treat a replica set member as a standalone server, bypassing auto-discovery. Other options include :direct, :replica_set, and :sharded.	Symbol	None
heartbeat_frequency	How often replica set members will communicate to check whether they are all alive.	Float	10
database	Database connection.	String	admin

Table 13.3 – MongoDB replica set client options configuration

Similar to connecting to a standalone server, there are also options for SSL and authentication that are used in the same way.

We can also configure the connection pool by setting the following code:

```
min_size,
max_size(setting this to 0 will create an unbounded connection
pool),
wait_queue_timeout(defaults to 1 in seconds).
```

The MongoDB driver will try to reuse existing connections, if available, or it will open a new connection. Once the pool limit has been reached, the driver will block further connections, waiting for a connection to be released to use it.

Replica set administration

The administration of a replica set can be significantly more complex than what is needed for single-server deployments. In this section, instead of trying to exhaustively cover all of the different cases, we will focus on some of the most common administrative tasks that we will have to perform and how to do them.

How to perform maintenance on replica sets

If we have some maintenance tasks that we have to perform in every member in a replica set, we always start with the secondaries. We perform maintenance by performing the following steps:

1. First, we connect to one of the secondaries via the mongo shell. Then, we stop that secondary as follows:

    ```
    > use admin
    > db.shutdownServer()
    ```

2. Then, using the same user that was connected to the mongo shell in the preceding step, we restart the mongo server as a standalone server in a different port as follows:

    ```
    > mongod --port 95658 --dbpath <wherever our mongoDB data
    resides in this host>
    ```

3. The following step is to connect to this mongod server (which is using dbpath) as follows:

    ```
    > mongo --port 37017
    ```

4. At this point, we can safely perform all of the administrative tasks on our standalone server without affecting our replica set operations. When we are done, we shut down the standalone server in the same way that we did in the first step.

5. We can then restart our server in the replica set by using the command line or the configuration script that we normally use. The final step is to verify that everything works fine by connecting to the replica set server and getting its replica set `status` as follows:

```
> rs.status()
```

The server should initially be in `state: RECOVERING`, and, once it has caught up with the secondary, it should be back in `state: SECONDARY`, like it was before starting the maintenance.

We need to repeat the same process for every secondary server. In the end, we have to perform maintenance on the primary. The only difference in the process for the primary is that we will start by stepping down our primary server into a secondary server before every other step, as follows:

```
> rs.stepDown(600)
```

By using the preceding argument, we prevent our secondary from being elected as a primary for 10 minutes. This should be enough time to shut down the server and continue with our maintenance as we did with the secondaries.

Initial sync

Adding a new member to an existing replica set requires copying all data from an existing member to the new member. MongoDB uses initial sync to copy all data in a first pass and then replication using the oplog to continuously keep the members in sync.

MongoDB offers logical initial sync and file copy-based initial sync options.

The logical initial sync will *not* clone the local database. It uses the oplog to sync from the primary, which means that we need to make sure that we have enough disk space in the target system to temporarily store the oplog records while the sync is ongoing.

File copy-based initial sync, on the other hand, is only available in the MongoDB Enterprise edition and copies the underlying database files directly between the source and target systems. As such, file copy-based initial sync will overwrite the local database in the target system. File copy-based initial sync can be faster than logical-based initial sync with a few limitations, primarily that it does not work at all with encrypted storage.

We can set the `initialSyncSourceReadPreference` parameter on `mongod` process startup to select between primary, secondary, or the nearest server as the source that MongoDB will try to sync from.

Resyncing a member of a replica set

Secondaries sync up with the primary by replaying the contents of the oplog. If our oplog is not large enough, or if we encounter network issues (partitioning, an underperforming network, or just an outage of the secondary server) for a period of time larger than the oplog, then MongoDB cannot use the oplog to catch up to the primary anymore.

At this point, we have the following three options:

- The more straightforward option is to delete our dbpath directory and restart the mongod process. In this case, MongoDB will start an initial sync from scratch. This option has the downside of putting a strain on our replica set and our network as well.

- The more complicated (from an operational standpoint) option is to copy data files from another well-behaving member of the replica set. This goes back to the contents of *Chapter 9, Monitoring, Backup, and Security*. The important thing to keep in mind is that a simple file copy will probably not suffice, as data files will have changed from the time that we started copying to the time that the copying ended.

- Finally, we can resync the member using the initial sync process as described in the preceding section.

Thus, we need to be able to take a snapshot copy of the filesystem under our data directory.

Another point of consideration is that by the time we start our secondary server with the newly copied files, our MongoDB secondary server will try to sync up to the primary using the oplog again. So, if our oplog has fallen so far behind the primary that it can't find the entry on our primary server, this method will fail, too.

> **Note**
> Keep a sufficiently sized oplog. Don't let data grow out of hand in any replica set member. Design, test, and deploy sharding early on.

Changing the oplog's size

Hand in hand with the preceding operational tip, we may need to rethink and resize our oplog as our data grows. Operations become more complicated and time-consuming as our data grows, and we need to adjust our oplog size to accommodate it. The steps for changing the oplog's size are as follows:

1. The first step is to restart our MongoDB secondary server as a standalone server, an operation that was described in the *How to perform maintenance on replica sets* section.

2. We then make a backup of our existing oplog as follows:

```
> mongodump --db local --collection 'oplog.rs' --port
37017
```

3. We keep a copy of this data, just in case. We then connect to our standalone database as follows:

```
> use local
> db = db.getSiblingDB('local')
> db.temp.drop()
```

Up until now, we have connected to the local database and deleted the temp collection, just in case it had any leftover documents.

4. The following step is to get the last entry of our current oplog and save it in the temp collection as follows:

```
> db.temp.save( db.oplog.rs.find( { }, { ts: 1, h: 1 }
).sort( {$natural : -1} ).limit(1).next() )
```

5. This following entry will be used when we restart our secondary server in order to track where it has reached in the oplog replication:

```
> db = db.getSiblingDB('local')
> db.oplog.rs.drop()
```

6. Now, we delete our existing oplog, and in the following step, we will create a new oplog of 4 GB in size as follows:

```
> db.runCommand( { create: "oplog.rs", capped: true,
size: (4 * 1024 * 1024 * 1024) } )
```

7. The following step is to copy the one entry from our temp collection back to our oplog as follows:

```
> db.oplog.rs.save( db.temp.findOne() )
```

8. Finally, we cleanly shut down our server from the admin database, using the db.shutdownServer() command, and restart our secondary as a member of the replica set.

9. We repeat this process for all secondary servers, and as a last step, we repeat the procedure for our primary member, which is done after we step the primary down by using the following command:

```
> rs.stepDown(600)
```

Reconfiguring a replica set when we have lost the majority of our servers

This is only intended as an interim solution and a last resort when we are faced with downtime and disrupted cluster operations. When we lose the majority of our servers and we still have enough servers to start a replica set (maybe including some quickly spawned arbiters), we can force a reconfiguration with only the surviving members.

First, we get the replica set configuration document as follows:

```
> cfg = rs.conf()
```

Using `printjson(cfg)`, we identify the members that are still operational. Let's say that these are 1, 2, and 3 as follows:

```
> cfg.members = [cfg.members[1] , cfg.members[2] , cfg.members[3]]
> rs.reconfig(cfg, {force : true})
```

By using `force : true`, we are forcing this reconfiguration to happen. Of course, we need to have at least three surviving members in our replica set for this to work.

> **Note**
>
> It's important to remove the failing servers as soon as possible by killing the processes and/or taking them out of the network to avoid unintended consequences; these servers may believe that they are still a part of a cluster that doesn't acknowledge them anymore.

Chained replication

Replication in MongoDB usually happens from the primary to the secondaries. In some cases, we may want to replicate from another secondary instead of the primary. Chained replication helps to alleviate the primary from read load, but at the same time, it increases the average replication lag for the secondary that chooses to replicate from a secondary. This makes sense, as replication has to go from the primary to the secondary (1), and then from this server to another secondary (2).

Chained replication can be enabled (and disabled, respectively) with the following `cfg` command:

```
> cfg.settings.chainingAllowed = true
```

In cases where `printjson(cfg)` doesn't reveal a `settings` subdocument, we need to create an empty one first, as follows:

```
> cfg.settings = { }
```

> **Note**
>
> If there is already a `settings` document, the preceding command will result in deleting its settings, leading to potential data loss.

Streaming replication

Starting from version 4.4, MongoDB is using streaming to replicate data from one server in a replica set to another. We can set `oplogFetcherUsesExhaust` to `false` to revert to the older, batched replication. This is only recommended if we have bandwidth constraints between the source and target or if streaming replication drains the source replication server.

Flow control

Starting from version 4.2, MongoDB is using flow control to keep lag for replica set writes that require a majority committed write concern under the configurable `flowControlTargetLagSeconds`, with a default value of 10 seconds.

Flow control limits the rate of writes to the primary to make sure secondaries can keep up with the primary writes. If the default value is not optimal, MongoDB recommends decreasing the value to force more aggressive flow control.

Cloud options for a replica set

We can set up and operate a replica set from our own servers, but we can reduce our operational overhead by using a **DBaaS** provider to do so. The three most widely used MongoDB cloud providers are Amazon's DocumentDB, Microsoft's Azure Cosmos DB, and MongoDB Atlas, the native offering from MongoDB, Inc.

In this section, we will go over these options and how they fare in comparison to using our own hardware and data centers.

Amazon DocumentDB

Amazon DocumentDB is a wire protocol-compatible MongoDB offering based on Amazon's infrastructure. It offers compatibility with versions 3.6 and 4.0 of MongoDB as of summer 2022.

Amazon DocumentDB is offering what is essentially a replica set configuration with one primary for writes and up to 15 secondaries across availability zones in a single region for reads. Sharding is not supported and write scaling can only be achieved by scaling up the primary server instance. We should accordingly scale up secondary instances to avoid replication lag in the case of high contention.

We can also have secondaries spanning multiple regions using global clusters. We can have up to five secondary clusters in different regions that replicate from the primary cluster and serve exclusively read operations. Any one of them can be promoted to be the primary cluster and start taking writes in the event of a region outage. Global clusters allow us to scale reads almost infinitely with typically up to 50 ms replication lag, allowing low-latency reads for users around the globe.

Amazon DocumentDB supports MongoDB ACID transactions as they were introduced in MongoDB version 4.0. It also supports implicit atomicity and consistency for CRUD operations (`findAndModify`, `update`, `insert`, and `delete`) that span multiple documents. In contrast, MongoDB supports atomicity and consistency for CRUD operations that span multiple subdocuments within one document.

Atomicity in bulk operations such as `updateMany` and `insertMany` is not guaranteed in the case of an error. This means that if, for example, `insertMany` attempts to insert 100 documents and encounters an error in one document, it will still insert the other 99 documents.

Amazon DocumentDB offers 99.9% SLA and high durability by storing six copies of each write across three availability zones in a region. There is no write concern concept in DocumentDB as all writes propagate to the majority of nodes before being acknowledged. The closest equivalent configuration in MongoDB for reference is $\{w: 3, j: true\}$.

More information on the Amazon DocumentDB SLA is available at `https://aws.amazon.com/documentdb/sla/`.

Amazon DocumentDB is not a drop-in replacement for MongoDB. Amazon has chosen to implement a subset of MongoDB features, leaving out, for example, MapReduce, capped collections, GridFS, and text and hashed indexes. If we are migrating from an existing MongoDB deployment, we need to make sure that our code is compatible with Amazon DocumentDB's feature list.

While DocumentDB can be a great choice within the AWS ecosystem, it cannot be considered a generic replacement for MongoDB Atlas.

Microsoft Azure Cosmos DB

Microsoft offers a **MongoDB Query Language (MQL)** compatible document-based storage option, Cosmos DB. Cosmos DB is compatible with MongoDB up to version 4.2 as of summer 2022.

Azure Cosmos DB offers a core API that is based on SQL and a number of extensions that offer Cassandra, graph database, or document-oriented database functionality.

Azure Cosmos DB's MongoDB API supports wire protocol compatibility with MongoDB versions 3.2, 3.6, 4.0, and 4.2. It is based on a serverless model; the user only needs to use the Azure connection string to connect to the database, taking away all management and operational maintenance tasks from the user.

Taking away operations using serverless also means that there is no concept of replication or sharding. Data is guaranteed to be available 99.999% of the time. Data durability within a single region is guaranteed, with the exception of an unrecoverable disaster in the region. Data durability across multiple regions depends on the consistency level configured at the account level.

Cosmos DB recommends the core (SQL) API for generic usage and the MongoDB API layer for migrations from MongoDB environments.

While Cosmos DB can be a great choice within the Azure ecosystem, it cannot be considered a generic replacement for MongoDB Atlas.

More information on the Microsoft Azure SLA is available at `https://azure.microsoft.com/en-us/support/legal/sla/cosmos-db/v1_4/`.

MongoDB Atlas

MongoDB Atlas is a newer offering from MongoDB, Inc., and it was launched in the summer of 2016. MongoDB Atlas offers the deployment of a replica set or sharded clusters in a single region or across multiple regions, all through a web interface or a command-line interface.

It always offers the latest MongoDB version, including quarterly rapid releases. If we want to always have the latest version of MongoDB with MongoDB, Inc. support, Atlas is the only option. If this is not a strict requirement, then we can wait for the annual major release (5.0, 6.0, and so on).

The only storage option is WiredTiger. There are multiple regions for each provider in the United States, Asia Pacific, Europe, Middle East, and Africa across the three biggest cloud platforms, AWS, Azure, and GCP. Most notably, the regions that are missing are the AWS China and AWS GovCloud US regions.

MongoDB Atlas also offers the serverless deployment option in public preview as of summer 2022. A serverless cluster is scaling reads and writes automatically and hides all of the operational complexity of a replica set or a sharded cluster from the MongoDB developer or administrator.

The running costs for cloud-hosted services can be significantly higher than setting them up on our own servers. What we gain in convenience and time to market may have to be repaid in operational costs.

We learned about our cloud hosting options for a replica set in this section. The following and final section will address the limitations of a replica set and when it is not enough to have a replica set topology on its own.

Replica set limitations

A replica set is great when we understand why we need it and what it cannot do. The different limitations for a replica set are as follows:

- It will not scale horizontally; we need sharding for it.

- We will introduce replication issues if our network is flaky.

- We will make debugging issues more complex if we use secondaries for reads, and these have fallen behind our primary server.

On the flip side, as we explained in earlier sections in this chapter, a replica set can be a great choice for replication, data redundancy, conforming with data privacy, backups, and even recovery from errors caused by humans or otherwise.

Summary

In this chapter, we discussed replica sets and how to administer them. Starting with an architectural overview of replica sets and replica set internals involving elections, we dove into setting up and configuring a replica set.

You learned how to perform various administrative tasks with replica sets, and you learned about the main options for outsourcing operations to a cloud DBaaS provider. Finally, we identified some of the limitations that replica sets in MongoDB currently have.

In the following chapter, we will move on to one of the most interesting concepts in MongoDB (which is used to achieve horizontal scaling): sharding.

14

Mastering Sharding

Sharding is the ability to horizontally scale out our database by partitioning our datasets across different servers—shards. This has been a feature of MongoDB since **version 1.6 (v1.6)** was released in August 2010. Foursquare and Bitly are two of MongoDB's most famous early customers and have used the sharding feature from its inception all the way to its general release.

In this chapter, we will learn about the following topics:

- How to design a sharding cluster and how to make the important decision of choosing the shard key

- Different sharding techniques and how to monitor and administrate sharded clusters

- The `mongos` router and how it is used to route our queries across different shards

- How we can recover from errors in our shards

By the end of this chapter, we will have mastered the underlying theory and concepts as well as the practical implementation of sharding using MongoDB.

This chapter covers the following topics:

- Why do we need sharding?

- Architectural overview

- Setting up sharding

- Sharding administration and monitoring

- Querying sharded data

- Sharding recovery

Technical requirements

To follow along with the code in this chapter, you need to connect to a MongoDB Atlas database. You can use the fully managed **database-as-a-service** (**DBaaS**) MongoDB Atlas offering, which provides a free tier as well as seamless upgrades to the latest version. You may need to switch to a paid plan to access the full range of sharding features on the MongoDB Atlas platform.

Why do we need sharding?

In database systems and computing systems in general, we have two ways to improve performance. The first one is to simply replace our servers with more powerful ones, keeping the same network topology and systems architecture. This is called **vertical scaling**.

An advantage of vertical scaling is that it is simple, from an operational standpoint, especially with cloud providers such as Amazon making it a matter of a few clicks to replace an `r6g.medium` server instance with an `r6g.extralarge` one. Another advantage is that we don't need to make any code changes, so there is little to no risk of something going catastrophically wrong.

The main disadvantage of vertical scaling is that there is a limit to it; we can only get servers that are as powerful as those our cloud provider can give to us.

A related disadvantage is that getting more powerful servers generally comes with an increase in cost that is not linear but exponential. So, even if our cloud provider offers more powerful instances, we will hit the cost-effectiveness barrier before we hit the limit of our department's credit card.

The second way to improve performance is by using the same servers with the same capacity and increasing their number. This is called **horizontal scaling**.

Horizontal scaling offers the advantage of theoretically being able to scale exponentially while remaining practical enough for real-world applications. The main disadvantage is that it can be operationally more complex and requires code changes and careful design of the system upfront. Horizontal scaling is also more complex when it comes to the system because it requires communication between the different servers over network links that are not as reliable as **inter-process communication** (**IPC**) on a single server. The following diagram shows the difference between horizontal and vertical scaling:

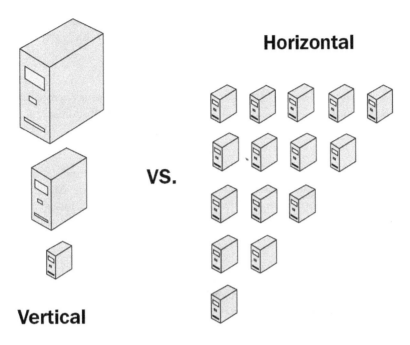

Figure 14.1: Conceptual architecture of vertical and horizontal scaling

To understand scaling, it's important to understand the limitations of single-server systems. A server is typically bound by one or more of the following characteristics:

- **Central processing unit (CPU)**: A CPU-bound system is one that is limited by our CPU's speed. A task such as the multiplication of matrices that can fit in **random-access memory (RAM)** will be CPU bound because there is a specific number of steps that have to be performed in the CPU without any disk or memory access needed for the task to complete. In this case, CPU usage is the metric that we need to keep track of.

- **Input/output (I/O)**: I/O-bound systems are similarly limited by the speed of our storage system **(hard-disk drive (HDD)** or **solid-state drive (SSD))**. A task such as reading large files from a disk to load into memory will be I/O bound as there is little to do in terms of CPU processing; a great majority of the time is spent reading files from the disk. The important metrics to keep track of are all metrics related to disk access, reads per second, and writes per second, compared to the practical limit of our storage system.

- **Memory and cache**: Memory-bound and cache-bound systems are restricted by the amount of available RAM and/or the cache size that we have assigned to them. A task that multiplies matrices larger than our RAM size will be memory bound, as it will need to page in/out data from the disk to perform the multiplication. The important metric to keep track of is the memory used. This may be misleading in MongoDB **Memory Mapped Storage Engine v1 (MMAPv1)**, as the storage engine will allocate as much memory as possible through the filesystem cache.

In the WiredTiger storage engine, on the other hand, if we don't allocate enough memory for the core MongoDB process, **out-of-memory (OOM)** errors may kill it, and this is something we want to avoid at all costs.

Monitoring memory usage has to be done both directly through the operating system and indirectly by keeping a track of page-in/out data. An increasing memory paging number is often an indication that we are running short of memory and the operating system is using virtual address space to keep up.

> **Note**
>
> MongoDB, being a database system, is generally memory and I/O bound. Investing in an SSD and more memory for our nodes is almost always a good investment. Most systems are a combination of one or more of the preceding limitations. Once we add more memory, our system may become CPU bound, as complex operations are almost always a combination of CPU, I/O, and memory usage.

MongoDB's sharding is simple enough to set up and operate, and this has contributed to its huge success over the years as it provides the advantages of horizontal scaling without requiring a large commitment of engineering and operations resources.

That being said, it's really important to get sharding right from the beginning, as it is extremely difficult from an operational standpoint to change the configuration once it has been set up. As we will learn in the following sections, sharding should not be an afterthought, but rather a key architectural design decision from an early point in the design process.

Architectural overview

A sharded cluster is comprised of the following elements:

- Two or more shards. Each shard must be a replica set.
- One or more query routers (`mongos`). A `mongos` router provides an interface between our application and the database.
- A replica set of config servers. Config servers store metadata and configuration settings for the entire cluster.

The relationships between these elements are shown in the following diagram:

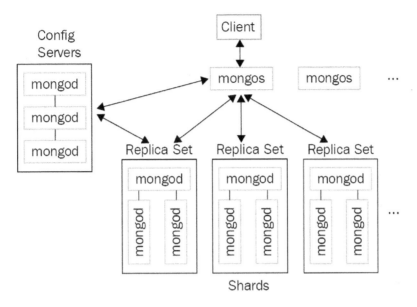

Figure 14.2: MongoDB sharded cluster architecture

Starting from MongoDB 3.6, shards must be implemented as replica sets.

Development, continuous deployment, and staging environments

In preproduction environments, it may be overkill to use the full set of servers. For efficiency reasons, we may opt to use a more simplified architecture.

The simplest possible configuration that we can deploy for sharding is the following:

- One mongos router
- One sharded replica set with one MongoDB server and two arbiters
- One replica set of config servers with three MongoDB servers

Support for multiple arbiters has been turned off by default in MongoDB 5.3. To enable it, we need to start each data server in the replica set with the allowMultipleArbiters=true parameter. Arbiters vote for successful operations but do not hold any data. So, for example, a write operation with write concern "majority" could succeed even when in our case we only have one data-bearing node and two arbiter nodes. Another potential issue is that when we change the replica set configuration using rs.reconfig() and we have multiple arbiters, the votes from multiple arbiters could elect as primary a node that has fallen behind in replication.

For these reasons, this configuration should only be used in development and staging environments and not in a production system.

A replica set of config servers must not have any delayed members or arbiters. All config servers must be configured to build indexes—this is enabled by default.

Staging is strongly recommended to mirror our production environment in terms of its servers, configuration, and (if possible) the stored dataset characteristics as well, in order to avoid surprises at deployment time.

Planning ahead with sharding

As we will see in the next sections, sharding is complicated and expensive, operation-wise. It is important to plan ahead and make sure that we start the sharding process long before we hit our system's limits.

Some rough guidelines on when you need to start sharding are provided here:

- When you have a CPU utilization of less than 70% on average
- When I/O (and, especially, write) capacity is less than 80%
- When memory utilization is less than 70% on average

As sharding helps with write performance, it's important to keep an eye on our I/O write capacity and the requirements of our application.

> **Note**
> Don't wait until the last minute to start sharding in an already busy-up-to-the-neck MongoDB system, as this can have unintended consequences.

Setting up sharding

Sharding is performed at the collection level. We can have collections that we don't want or need to shard for several reasons. We can leave these collections unsharded.

These collections will be stored in the primary shard. The primary shard is different for each database in MongoDB. The primary shard is automatically selected by MongoDB when we create a new database in a sharded environment. MongoDB will pick the shard that has the least data stored at the moment of creation.

If we want to change the primary shard at any other point, we can issue the following command:

```
> db.runCommand( { movePrimary : "mongo_books", to : "UK_based"
} )
```

With this, we move the database named mongo_books to the shard named UK_based.

Choosing the shard key

Choosing our shard key is the most important decision we need to make: once we shard our data and deploy our cluster, it becomes very difficult to change the shard key. First, we will go through the process of changing the shard key.

Changing the shard key prior to v4.2

Prior to MongoDB v4.2, there was no command or simple procedure to change the shard key in MongoDB. The only way to change the shard key involved backing up and restoring all of our data, something that may range from being extremely difficult to impossible in high-load production environments.

Here are the steps that we need to go through in order to change the shard key:

1. Export all data from MongoDB.

2. Drop the original sharded collection.

3. Configure sharding with the new key.

4. Pre-split the new shard key range.

5. Restore our data back into MongoDB.

Of these steps, *step 4* is the one that needs further explanation.

MongoDB uses chunks to split data in a sharded collection. If we bootstrap a MongoDB sharded cluster from scratch, chunks will be calculated automatically by MongoDB. MongoDB will then distribute the chunks across different shards to ensure that there is an equal number of chunks in each shard.

The only time when we cannot really do this is when we want to load data into a newly sharded collection.

The reasons for this are threefold, as explained here:

- MongoDB creates splits only after an `insert` operation.

- Chunk migration will copy all of the data in that chunk from one shard to another.

- `floor(n/2)` chunk migrations can happen at any given time, where n is the number of shards we have. Even with three shards, this is only a `floor(1.5)=1` chunk migration at a time.

These three limitations mean that letting MongoDB figure it out on its own will definitely take much longer, and may result in an eventual failure. This is why we want to pre-split our data and give MongoDB some guidance on where our chunks should go.

In our example of the `mongo_books` database and the `books` collection, this is how it would look:

```
> db.runCommand( { split : "mongo_books.books", middle : { id : 50 } } )
```

The `middle` command parameter will split our key space into documents that have an `id` value less than or equal to `50` and documents that have an `id` value greater than `50`. There is no need for a document to exist in our collection with an `id` value that is equal to `50` as this will only serve as the guidance value for our partitions.

In this example, we chose `50`, as we assume that our keys follow a uniform distribution (that is, there is the same count of keys for each value) in the range of values from `0` to `100`.

> **Note**
>
> We should aim to create at least 20-30 chunks to grant MongoDB flexibility in potential migrations. We can also use `bounds` and `find` instead of `middle` if we want to manually define the partition key, but both parameters need data to exist in our collection before applying them.

Changing the shard key – MongoDB v5, v6, and beyond

MongoDB v4.2 allowed users to modify a shard key's value. MongoDB v4.4 allowed users to add a suffix to an existing shard key.

MongoDB v5.0 introduced live resharding, allowing users to change the shard key using a simple administrative command, as shown here:

```
sh.reshardCollection("<database>.<collection>", "<new_
shardkey>")
```

We need to have at least 1.2 times the storage size of the collection for this operation to succeed. For example, if we reshard a collection that is 100 **gigabytes** (**GB**) in size, we need to make sure that we have at least 120 GB of free space across the cluster.

MongoDB recommends that CPU usage is less than 80% and I/O usage less than 50% before we attempt to reshard a collection.

Our application should be able to tolerate writes waiting for up to 2 seconds before being acknowledged while the resharding operation is in progress.

No index builds should be in progress before invoking the resharding command.

We should be cautious before invoking data definition-altering commands while resharding is in progress—these include, for example, index maintenance and collection definition changes. For example, changing the collection name while resharding this collection is not allowed, for good reasons.

Application code should be able to use both the current shard key and the new shard key if we want to ensure application availability while the resharding process is in progress.

Changing the shard key is administratively easy, but operationally heavy. Good shard key design is still needed, and that's what we are going to learn how to do in the next section.

Choosing the correct shard key

After the previous section, it's now self-evident that we need to take the choice of our shard key into consideration as this is a decision that we have to stick with.

A great shard key has the following three characteristics:

- High cardinality
- Low frequency
- Nonmonotonically changing values

We will go over the definitions of these three properties first to understand what they mean, as follows:

- **High cardinality**: This means that the shard key must have as many distinct values as possible. A Boolean can take only the values of `true`/`false`, and so it is a bad shard-key choice. A 64-bit long value field that can take any value from $-(2^{63})$ to $2^{63}-1$ is a good shard-key choice, in terms of cardinality.

- **Low frequency**: This directly relates to the argument about high cardinality. A low-frequency shard key will have a distribution of values as close to a perfectly random/uniform distribution. Using the example of our 64-bit long value, it is of little use to us if we have a field that can take values ranging from $-(2^{63})$ to $2^{63}-1$ if we end up observing the values of 0 and 1 all the time. In fact, it is as bad as using a Boolean field, which can also take only two values. If we have a shard key with high-frequency values, we will end up with chunks that are indivisible. These chunks cannot be further divided and will grow in size, negatively affecting the performance of the shard that contains them.

- **Nonmonotonically changing values**: This means that our shard key should not be, for example, an integer that always increases with every new insert. If we choose a monotonically increasing value as our shard key, this will result in all writes ending up in the last of all of our shards, limiting our write performance.

> **Note**
> If we want to use a monotonically changing value as the shard key, we should consider using hash-based sharding.

In the next section, we will describe different sharding strategies, including their advantages and disadvantages.

Range-based sharding

The default and most widely used sharding strategy is range-based sharding. This strategy will split our collection's data into chunks, grouping documents with nearby values in the same shard.

For our example database and collection (`mongo_books` and `books`, respectively), we have the following:

```
> sh.shardCollection("mongo_books.books", { id: 1 } )
```

This creates a range-based shard key on `id` with an ascending direction. The direction of our shard key will determine which documents will end up in the first shard and which ones will appear in subsequent ones.

This is a good strategy if we plan to have range-based queries, as these will be directed to the shard that holds the result set instead of having to query all shards.

Hash-based sharding

If we don't have a shard key (or can't create one) that achieves the three goals mentioned previously, we can use the alternative strategy of hash-based sharding. In this case, we are trading data distribution with query isolation.

Hash-based sharding will take the values of our shard key and hash them in a way that guarantees close to uniform distribution. This way, we can be sure that our data will be evenly distributed across the shards. The downside is that only exact match queries will get routed to the exact shard that holds the value. Any range query will have to go out and fetch data from all the shards.

For our example database and collection (`mongo_books` and `books`, respectively), we have the following:

```
> sh.shardCollection("mongo_books.books", { id: "hashed" } )
```

Similar to the preceding example, we are now using the `id` field as our hashed shard key.

Suppose we use fields with float values for hash-based sharding. Then, we will end up with collisions if the precision of our floats is more than 2^{53}. These fields should be avoided where possible.

Hash-based sharding can also use a compound index with a single hashed field. Creating such an index is supported since v4.4, like so:

```
db['mongo_books'].books.createIndex( { "price" : 1, "isbn" :
-1, "id" : "hashed" } )
```

This will create a compound hashed index on `price` ASC and `isbn` DESC, and `id` as the hashed field. We can use the compound hashed index for sharding in a few different ways. We can prefix the index with the hashed field to resolve data distribution issues with monotonically increasing fields.

On the other hand, we can implement zone-based sharding by prefixing the compound hashed index with the fields that we will use to split our data into different chunks and suffix the index with the hashed field to achieve a more even distribution of sharded data.

Coming up with our own key

Range-based sharding does not need to be confined to a single key. In fact, in most cases, we would like to combine multiple keys to achieve high cardinality and low frequency.

A common pattern is to combine a low-cardinality first part (but still with a number of distinct values more than two times the number of shards that we have) with a high-cardinality key as its second field. This achieves both read and write distribution from the first part of the sharding key and then cardinality and read locality from the second part.

On the other hand, if we don't have range queries, then we can get away with using hash-based sharding on a **primary key (PK)**, as this will exactly target the shard and document that we are going after.

To make things more complicated, these considerations may change depending on our workload. A workload that consists almost exclusively (say, 99.5%) of reads won't care about write distribution. We can use the built-in `_id` field as our shard key, and this will only add 0.5% load to the last shard. Our reads will still be distributed across shards. Unfortunately, in most cases, this is not simple.

Location-based data

Because of government regulations and the desire to have our data as close to our users as possible, there is often a constraint and need to limit data in a specific data center. By placing different shards at different data centers, we can satisfy this requirement.

> **Note**
>
> Every shard is essentially a replica set. We can connect to it as we would connect to a replica set for administrative and maintenance operations. We can query one shard's data directly, but the results will only be a subset of the full sharded result set.

After learning how to start a sharded cluster, we will move on to discuss the administration and monitoring of sharded clusters.

Sharding administration and monitoring

Sharded MongoDB environments have some unique challenges and limitations compared to single-server or replica-set deployments. In this section, we will explore how MongoDB balances our data across shards using chunks and how we can tweak them if we need to. Together, we will explore some of sharding's design limitations.

Balancing data – how to track and keep our data balanced

One of the advantages of sharding in MongoDB is that it is mostly transparent to the application and requires minimal administration and operational effort.

One of the core tasks that MongoDB needs to perform continuously is balancing data between shards. No matter whether we implement range-based or hash-based sharding, MongoDB will need to calculate bounds for the hashed field to be able to figure out which shard to direct every new document insert or update toward. As our data grows, these bounds may need to get readjusted to avoid having a hot shard that ends up with the majority of our data.

For the sake of this example, let's assume that there is a data type named `extra_tiny_int` with integer values from [-12, 12). If we enable sharding on this `extra_tiny_int` field, then the initial bounds of our data will be the whole range of values denoted by `$minKey: -12` and `$maxKey: 11`.

After we insert some initial data, MongoDB will generate chunks and recalculate the bounds of each chunk to try to balance our data.

> **Note**
>
> By default, the initial number of chunks created by MongoDB is *2 × the number of shards*.

In our case of two shards and four initial chunks, the initial bounds will be calculated as follows:

Chunk1: [-12..-6)

Chunk2: [-6..0)

Chunk3: [0..6)

Chunk4: [6,12) where [is inclusive and) is not inclusive

The following diagram illustrates the preceding explanation:

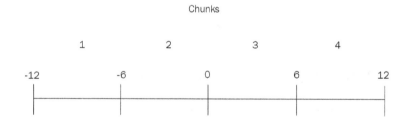

Figure 14.3: Sharding chunks and boundaries

After we insert some data, our chunks will look like this:

- *ShardA:*

  - *Chunk1: -12,-8,-7*

  - *Chunk2: -6*

- *ShardB:*

  - *Chunk3: 0, 2*

  - *Chunk4: 7, 8, 9, 10, 11, 11, 11, 11*

The following diagram illustrates the preceding explanation:

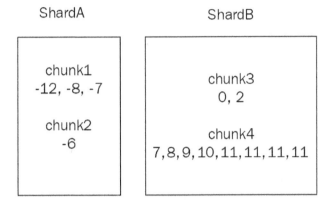

Figure 14.4: Sharding chunks and shard allocation

In this case, we observe that *chunk4* has more items than any other chunk. MongoDB will first split *chunk4* into two new chunks, attempting to keep the size of each chunk under a certain threshold (128 **megabytes (MB)** by default, starting from MongoDB 6.0).

Now, instead of *chunk4*, we have *chunk4A: 7, 8, 9, 10* and *chunk4B: 11, 11, 11, 11*.

The following diagram illustrates the preceding explanation:

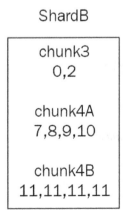

Figure 14.5: Sharding chunks and shard allocation (continued)

The new bounds of this are provided here:

- *chunk4A: [6,11)*
- *chunk4B: [11,12)*

Note that *chunk4B* can only hold one value. This is now an indivisible chunk—a chunk that cannot be broken down into smaller ones anymore—and will grow in size unbounded, causing potential performance issues down the line.

This clarifies why we need to use a high-cardinality field as our shard key and why something such as a Boolean, which only has `true/false` values, is a bad choice for a shard key.

In our case, we now have two chunks in *ShardA* and three chunks in *ShardB*. Let's look at the following table:

Number of chunks	Migration threshold
≤19	2
20-79	4
≥80	8

Table 14.1: Sharding number of chunks in each shard and migration threshold

We have not reached our migration threshold yet, since *3-2 = 1*.

The migration threshold is calculated as the number of chunks in the shard with the highest count of chunks and the number of chunks in the shard with the lowest count of chunks, as follows:

- *Shard1 -> 85 chunks*
- *Shard2 -> 86 chunks*
- *Shard3 -> 92 chunks*

In the preceding example, balancing will not occur until *Shard3* (or *Shard2*) reaches *93* chunks because the migration threshold is *8* for *≥80* chunks, and the difference between *Shard1* and *Shard3* is still *7* chunks (*92-85*).

If we continue adding data in *chunk4A*, it will eventually be split into *chunk4A1* and *chunk4A2*.

We now have four chunks in *ShardB* (*chunk3*, *chunk4A1*, *chunk4A2*, and *chunk4B*) and two chunks in *ShardA* (*chunk1* and *chunk2*).

The following diagram illustrates the relationships of the chunks to the shards:

ShardB

Figure 14.6: Sharding chunks and shards

The MongoDB balancer will now migrate one chunk from *ShardB* to *ShardA* as *4-2 = 2*, reaching the migration threshold for fewer than *20* chunks. The balancer will adjust the boundaries between the two shards in order to be able to query more effectively (targeted queries). The following diagram illustrates the preceding explanation:

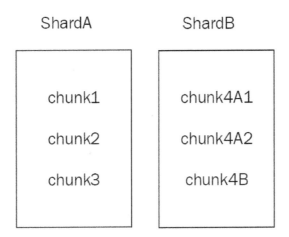

Figure 14.7: Sharding chunks and shards (continued)

As you can see from the preceding diagram, MongoDB will try to split >*128* MB chunks in half in terms of size. The bounds between the two resulting chunks may be completely uneven if our data distribution is uneven to begin with. MongoDB can split chunks into smaller ones but cannot merge them automatically. We need to manually merge chunks—a delicate and operationally expensive procedure.

Chunk administration

Most of the time, we should leave chunk administration to MongoDB. We should manually manage chunks at the start, upon receiving the initial load of data, when we change our configuration from a replica set to sharding.

Global shard read/write concern

We can set the read and write concern in a replica-set level via the primary server's mongod server. We can set the read and write concern in a sharded cluster level via the mongos server.

We need to run the following command while connected to the admin database:

```
db.adminCommand(
    {
        setDefaultRWConcern : 1,
        defaultReadConcern: { <read concern> },
        defaultWriteConcern: { <write concern> },
        writeConcern: { <write concern> },
        comment: <any>
    }
)
```

This contains the following parameters:

- setDefaultRWConcern must be set to 1 (true).

- Either defaultReadConcern or defaultWriteConcern or both of these parameters need to be set.

- The defaultReadConcern parameter for the read concern can only be one of the local, available, or majority levels.

- The defaultWriteConcern parameter for the write concern can be any w value greater than 0. This means that at least one server needs to acknowledge the write.

- The writeConcern parameter is the write concern to use for this command.

- comment is an optional description that will appear in logs and other mongo auditing storage.

Moving chunks

To move a chunk manually, we need to issue the following command after connecting to mongos and the admin database:

```
> db.runCommand( { moveChunk : 'mongo_books.books' ,
  find : {id: 50},
  to : 'shard1.packtdb.com' } )
```

Using the preceding command, we move the chunk containing the document with id: 50 (this has to be the shard key) from the books collection of the mongo_books database to the new shard named shard1.packtdb.com.

We can also more explicitly define the bounds of the chunk that we want to move. Now, the syntax looks like this:

```
> db.runCommand( { moveChunk : 'mongo_books.books' ,
  bounds :[ { id : <minValue> } ,
  { id : <maxValue> } ],
  to : 'shard1.packtdb.com' } )
```

Here, minValue and maxValue are the values that we get from db.printShardingStatus().

In the example used previously, for *chunk2*, minValue would be -6 and maxValue would be 0.

> **Note**
>
> Do not use find in hash-based sharding. Use bounds instead.

Changing the default chunk size

To change the default chunk size, we need to connect to a mongos router and—consequently—to the config database.

Then, we issue the following command to change our global chunksize value to 16 MB:

```
> db.settings.save( { _id:"chunksize", value: 16 } )
```

The main reasoning behind changing chunksize comes from cases where the default chunksize value of 128 MB can cause more I/O than our hardware can handle. In this case, defining a smaller chunksize value will result in more frequent but less data-intensive migrations.

Changing the default chunk size has the following drawbacks:

- Creating more splits by defining a smaller chunk size cannot be undone automatically.

- Increasing the chunk size will not force any chunk migration; instead, chunks will grow through inserts and updates until they reach the new size.

- Lowering the chunk size may take quite some time to complete.

- Automatic splitting to comply with the new chunk size if it is lower will only happen upon an insert or update operation. We may have chunks that don't get any write operations and thus will not be changed in size.

The chunk size can be from 1 to 1,024 MB.

Jumbo chunks

In rare cases, we may end up with jumbo chunks—chunks that are larger than the chunk size and cannot be split by MongoDB. We may also run into the same situation if the number of documents in our chunk exceeds the maximum document limit.

These chunks will have the jumbo flag enabled. Ideally, MongoDB will keep track of whether it can split the chunk, and, as soon as it can, it will get split; however, we may decide that we want to manually trigger a split before MongoDB does.

The way to do this is set out in the following steps:

1. Connect via a shell to your mongos router and run the following code:

   ```
   > sh.status(true)
   ```

2. Identify the chunk that has jumbo in its description using the following code:

   ```
   databases:
   ...
   mongo_books.books
   . . .
   chunks:
   ...
     shardB 2
     shardA 2
     { "id" : 7 } -->> { "id" : 9 } on : shardA Timestamp(2,
   2) jumbo
   ```

3. Invoke splitAt() or splitFind() manually to split the chunk on the books collection of the mongo_books database at the id value that is equal to 8 using the following code:

```
> sh.splitAt( "mongo_books.books", { id: 8 })
```

> **Note**
>
> The splitAt() function will split based on the split point we define. The two new splits may or may not be balanced.

Alternatively, if we want to leave it to MongoDB to find where to split our chunk, we can use splitFind, as follows:

```
> sh.splitFind("mongo_books.books", {id: 7})
```

The splitFind phrase will try to find the chunk that the id: 7 query belongs to and automatically define new bounds for the split chunks so that they are roughly balanced.

In both cases, MongoDB will try to split the chunk, and if successful, it will remove the jumbo flag from it.

4. If the preceding operation is unsuccessful, then—and only then—should we try stopping the balancer first, while also verifying the output and waiting for any pending migrations to finish first, as shown in the following code snippet:

```
> sh.stopBalancer()
> sh.getBalancerState()
> use config
while( sh.isBalancerRunning() ) {
  print("waiting...");
  sleep(1000);
}
```

This should return false.

5. Wait for any waiting... messages to stop printing, and then find the jumbo-flagged chunk in the same way as before.

6. Then, update the chunks collection in your config database of the mongos router, like this:

```
> db.getSiblingDB("config").chunks.update(
{ ns: "mongo_books.books", min: { id: 7 }, jumbo: true
},
{ $unset: { jumbo: "" } }
)
```

The preceding command is a regular `update()` command, with the first argument being the `find()` part to find out which document to update and the second argument being the operation to apply to it (`$unset: jumbo flag`).

7. After all this is done, we re-enable the balancer, as follows:

```
> sh.setBalancerState(true)
```

8. Then, we connect to the `admin` database to flush the new configuration to all nodes, as follows:

```
> db.adminCommand({ flushRouterConfig: 1 } )
```

> **Note**
>
> Always back up the `config` database before modifying any state manually.

Merging chunks

As we have seen previously, MongoDB will usually adjust the bounds for each chunk in our shard to make sure that our data is equally distributed. This may not work in some cases—especially when we define chunks manually—if our data distribution is surprisingly unbalanced, or if we have many `delete` operations in our shard.

Having empty chunks will invoke unnecessary chunk migrations and give MongoDB a false impression of which chunk needs to be migrated. As we have explained before, the threshold for chunk migration is dependent on the number of chunks that each shard holds. Having empty chunks may or may not trigger the balancer when it's needed.

Chunk merging can only happen when at least one of the chunks is empty, and only between adjacent chunks.

To find empty chunks, we need to connect to the database that we want to inspect (in our case, `mongo_books`) and use `runCommand` with `dataSize` set, as follows:

```
> use mongo_books
> db.runCommand({
  "dataSize": "mongo_books.books",
  "keyPattern": { id: 1 },
  "min": { "id": -6 },
  "max": { "id": 0 }
})
```

The `dataSize` phrase follows the `database_name.collection_name` pattern, whereas `keyPattern` is the shard key that we have defined for this collection.

The `min` and `max` values should be calculated by the chunks that we have in this collection. In our case, we have entered *chunkB*'s details from the example earlier in this chapter.

If the bounds of our query (which, in our case, are the bounds of *chunkB*) return no documents, the result will resemble the following:

```
{ "size" : 0, "numObjects" : 0, "millis" : 0, "ok" : 1 }
```

Now that we know that *chunkB* has no data, we can merge it with another chunk (in our case, this could only be *chunkA*), like so:

```
> db.runCommand( { mergeChunks: "mongo_books.books",
  bounds: [ { "id": -12 },
  { id: 0 } ]
  } )
```

Upon success, this will return MongoDB's default `ok` status message, as shown in the following code snippet:

```
{ "ok" : 1 }
```

We can then verify that we only have one chunk on *ShardA* by invoking `sh.status()` again.

Adding and removing shards

Adding a new shard to our cluster is as easy as connecting to `mongos`, connecting to the `admin` database, and invoking `runCommand` with the following code:

```
> db.runCommand( {
addShard: "mongo_books_replica_set/rs01.packtdb.com:27017",
maxSize: 18000, name: "packt_mongo_shard_UK"
} )
```

This adds a new shard from the replica set named `mongo_books_replica_set` from the `rs01.packtdb.com` host running on port `27017`. We also define the `maxSize` value of data for this shard as `18000` MB (or we can set it to `0` to give it no limit) and the name of the new shard as `packt_mongo_shard_UK`.

> **Note**
>
> This operation will take quite some time to complete as chunks will have to be rebalanced and migrated to the new shard.

Removing a shard, on the other hand, requires more involvement since we have to make sure that we won't lose any data on the way. We do this like so:

1. First, we need to make sure that the balancer is enabled using `sh.getBalancerState()`. Then, after identifying the shard we want to remove using any one of the `sh.status()`, `db.printShardingStatus()`, or `listShards` admin commands, we connect to the `admin` database and invoke `removeShard`, as follows:

    ```
    > use admin
    > db.runCommand( { removeShard: "packt_mongo_shard_UK" }
    )
    ```

    The output should contain the following elements:

    ```
    . . .
      "msg" : "draining started successfully",
      "state" : "started",
    . . .
    ```

2. Then, if we invoke the same command again, we get the following output:

    ```
    > db.runCommand( { removeShard: "packt_mongo_shard_UK" }
    )

    ...
    "msg" : "draining ongoing",
     "state" : "ongoing",
     "remaining" : {
     "chunks" : NumberLong(2),
     "dbs" : NumberLong(3)
     },
    ...
    ```

    The remaining document in the result contains the number of `chunks` and `dbs` instances that are still being transferred. In our case, it's 2 and 3 respectively.

> **Note**
>
> All the commands need to be executed in the `admin` database.

An extra complication in removing a shard can arise if the shard we want to remove serves as the primary shard for one or more of the databases that it contains. The primary shard is allocated by MongoDB when we initiate sharding, so when we remove the shard, we need to manually move these databases to a new shard.

3. We will know whether we need to perform this operation by looking at the following section of the result from `removeShard()`:

```
...
"note" : "you need to drop or movePrimary these
databases",
  "dbsToMove" : [
  "mongo_books"
  ],
...
```

We need to drop or perform `movePrimary` on our `mongo_books` database. The way to do this is to first make sure that we are connected to the `admin` database.

> **Note**
>
> We need to wait for all of the chunks to finish migrating before running the previous command.

4. Make sure that the result contains the following elements before proceeding:

```
..."remaining" : {
"chunks" : NumberLong(0) }...
```

5. Only after we have made sure that the chunks to be moved are down to zero can we safely run the following command:

```
> db.runCommand( { movePrimary: "mongo_books", to:
"packt_mongo_shard_EU" })
```

6. This command will invoke a blocking operation, and when it returns, it should have the following result:

```
{ "primary" : "packt_mongo_shard_EU", "ok" : 1 }
```

7. Invoking the same `removeShard()` command after we are all done should return the following result:

```
> db.runCommand( { removeShard: "packt_mongo_shard_UK" }
)

... "msg" : "removeshard completed successfully",
  "state" : "completed",
  "shard" : "packt_mongo_shard_UK"
```

```
"ok" : 1
...
```

8. Once we get to `state` as `completed` and `ok` as `1`, it is safe to remove our `packt_mongo_shard_UK` shard.

Removing a shard is naturally more complicated than adding one. We need to allow some time, hope for the best, and plan for the worst when performing potentially destructive operations on our live cluster.

Sharding limitations

Sharding comes with great flexibility. Unfortunately, there are a few limitations in the way that we perform some of the operations.

We will highlight the most important ones in the following list:

- The `group()` database command does not work. The `group()` command should not be used anyway; use `aggregate()` and the aggregation framework instead, or `mapreduce()`.

- The `db.eval()` command does not work and should be disabled in most cases for security reasons.

- The `$isolated` option for updates does not work. This is a functionality that is missing in sharded environments. The `$isolated` option for `update()` provides the guarantee that, if we update multiple documents at once, other readers and writers will not see some of the documents updated with the new value, and the others will still have the old value. The way this is implemented in unsharded environments is by holding a global write lock and/or serializing operations to a single thread to make sure that every request for the documents affected by `update()` will not be accessed by other threads/operations. This implementation means that it is not performant and does not support any concurrency, which makes it prohibitive to allow the `$isolated` operator in a sharded environment.

- The `$snapshot` operator for queries is not supported. The `$snapshot` operator in the `find()` cursor prevents documents from appearing more than once in the results, as a result of being moved to a different location on the disk after an update. The `$snapshot` operator is operationally expensive and often not a hard requirement. The way to substitute it is by using an index for our queries on a field whose keys will not change for the duration of the query.

- The indexes cannot cover our queries if our queries do not contain the shard key. Results in sharded environments will come from the disk and not exclusively from the index. The only exception is if we query only on the built-in `_id` field and return only the `_id` value, in which case, MongoDB can still cover the query using built-in indexes.

- update() and remove() operations work differently. All update() and remove() operations in a sharded environment must include either the _id field of the documents that are to be affected or the shard key; otherwise, the mongos router will have to do a full table scan across all collections, databases, and shards, which would be operationally very expensive.

- Unique indexes across shards need to contain the shard key as a prefix of the index. In other words, to achieve the uniqueness of documents across shards, we need to follow the data distribution that MongoDB follows for the shards.

- The shard key has to be up to 512 bytes in size. The shard key index has to be in ascending order on the key field that gets sharded and—optionally—other fields as well, or a hashed index on it.

After discussing sharding administration, monitoring, and its limitations, we will move on to learn how it can be different querying data stored in a sharded cluster as opposed to a single-server or a single-replica-set architecture.

Querying sharded data

Querying our data using a MongoDB shard is different than a single-server deployment or a replica set. Instead of connecting to the single server or the primary of the replica set, we connect to the mongos router, which decides which shard to ask for our data. In this section, we will explore how the query router operates and use Ruby to illustrate how similar to a replica set this is for the developer.

The query router

The query router, also known as the mongos process, acts as the interface and entry point to our MongoDB cluster. Applications connect to it instead of connecting to the underlying shards and replica sets; mongos executes queries, gathers results, and passes them to our application.

The mongos process doesn't hold any persistent state and is typically low on system resources. It acts as a proxy for requests. When a query comes in, mongos will examine it, decide which shards need to execute the query, and establish a cursor in each one of them.

> **Note**
>
> The mongos process is typically hosted in the same instance as the application server.

find

If our query includes the shard key or a prefix of the shard key, mongos will perform a targeted operation, only querying the shards that hold the keys we are looking for.

For example, with a composite shard key of {_id, email, address} on our User collection, we can have a targeted operation with any of the following queries:

```
> db.User.find({_id: 1})
> db.User.find({_id: 1, email: 'alex@packt.com'})
> db.User.find({_id: 1, email: 'janluc@packt.com', address:
'Linwood Dunn'})
```

These queries consist of either a prefix (as is the case with the first two) or the complete shard key.

On the other hand, a query on {email, address} or {address} will not be able to target the right shards, resulting in a broadcast operation. A broadcast operation is any operation that doesn't include the shard key or a prefix of the shard key, and they result in mongos querying every shard and gathering results from them. They are also known as **scatter-and-gather operations** or **fan-out queries**.

> **Note**
>
> This behavior is a direct result of the way indexes are organized and is similar to the behavior that we identified in the chapter about indexing.

Query results may be affected by transaction writes. For example, if we have multiple shards and a transaction writes document A in shard A, which is committed, and document B in shard B, which is not yet committed, querying with read concern local in shard A will return document A, ignoring document B in shard B. Because of the read concern "local", MongoDB will not reach out into shard B to retrieve document B.

sort/limit/skip

If we want to sort our results, we have the following two options:

- If we are using the shard key in our sort criteria, then mongos can determine the order in which it has to query a shard or shards. This results in an efficient and, again, targeted operation.

- If we are not using the shard key in our sort criteria, then—as is the case with a query without any sort criteria—it's going to be a fan-out query. To sort the results when we are not using the shard key, the primary shard executes a distributed merge sort locally before passing on the sorted result set to mongos.

A limit on the queries is enforced on each individual shard and then again at the mongos level, as there may be results from multiple shards. A skip operator, on the other hand, cannot be passed on to individual shards and will be applied by mongos after retrieving all the results locally.

If we combine the `skip` and `limit` operators, `mongos` will optimize the query by passing both values to individual shards. This is particularly useful in cases such as pagination. If we query without `sort` and the results are coming from more than one shard, `mongos` will round-robin across shards for the results.

update/remove

In document modifier operations, such as `update` and `remove`, we have a similar situation to the one we saw with `find`. If we have the shard key in the `find` section of the modifier, then `mongos` can direct the query to the relevant shard.

If we don't have the shard key in the `find` section, then it will again be a fan-out operation.

> **Note**
>
> `UpdateOne`, `replaceOne`, and `removeOne` operations must have the shard key or the `_id` value.

The following table sums up the operations that we can use with sharding:

Type of operation	Query topology
Insert	Must have the shard key
Update	Can have the shard key
Query with shard key	Targeted operation
Query without shard key	Scatter-and-gather operation/fan-out query
Indexed, sorted query with shard key	Targeted operation
Indexed, sorted query without shard key	Distributed sort merge

Table 14.2: Sharding and create, read, update, and delete (CRUD) operations

Hedged reads

If we use non-primary read preference and MongoDB v>=4.4, then the MongoDB router (`mongos`) will by default attempt to hedge reads—for example, when we invoke `find()`. For each operation, it will send the request to two members for each shard (replica set). Whichever request returns first will be the return value, and the other request will either return and be discarded or time out after `maxTimeMSForHedgedReads` **milliseconds (msecs)**—the default is 150.

Querying using Ruby

Connecting to a sharded cluster using Ruby or any other language is no different than connecting to a replica set. Using the official Ruby driver, we have to configure the `client` object to define a set of `mongos` servers, as shown in the following code snippet:

```
client = Mongo::Client.new('mongodb://key:password@mongos-
server1-host:mongos-server1-port,mongos-server2-host:mongos-
server2-port/admin?ssl=true&authSource=admin')
```

`mongo-ruby-driver` will then return a `client` object, which is no different than connecting to a replica set from the Mongo Ruby client. We can then use the `client` object as we did in previous chapters, with all the caveats around how sharding behaves differently from a standalone server or a replica set with regard to querying and performance.

Performance comparison with replica sets

Developers and architects are always looking out for ways to compare performance between replica sets and sharded configurations.

The way MongoDB implements sharding is based on top of replica sets. Every shard in production should be a replica set. The main difference in performance comes from fan-out queries. When we are querying without the shard key, MongoDB's execution time is limited by the worst-performing replica set. In addition, when using sorting without the shard key, the primary server has to implement the distributed merge sort on the entire dataset. This means that it has to collect all data from different shards, merge sort them, and pass them as sorted to `mongos`. In both cases, network latency and limitations in bandwidth can slow down operations, which they wouldn't do with a replica set.

On the flip side, by having three shards, we can distribute our working-set requirements across different nodes, thereby serving results from RAM instead of reaching out to the underlying storage—HDD or SSD.

On the other hand, writes can be sped up significantly since we are no longer bound by a single node's I/O capacity, and we can have writes in as many nodes as there are shards. Summing up, in most cases—and especially for cases where we are using the shard key—both queries and modification operations will be significantly sped up by sharding.

> **Note**
> The shard key is the single most important decision in sharding and should reflect and apply to our most common application use cases.

Querying a sharded cluster needs some extra consideration from the application developer's perspective. In the next section, we will learn how to recover from operational issues.

Sharding recovery

In this section, we will explore different failure types and how we can recover in a sharded environment. Failure in a distributed system can take multiple forms and shapes. In this section, we will cover all possible failure cases, as outlined here:

- A mongos process breaks
- A mongod process breaks
- A config server breaks
- A shard goes down
- The entire cluster goes down

In the following sections, we will describe how each component failure affects our cluster and how to recover.

mongos

The mongos process is a relatively lightweight process that holds no state. In the case that the process fails, we can just restart it or spin up a new process on a different server. It's recommended that mongos processes are located in the same server as our application, and so it makes sense to connect from our application using the set of mongos servers that we have colocated in our application servers to ensure **high availability (HA)** of mongos processes.

mongod

A mongod process failing in a sharded environment is no different than it failing in a replica set. If it is a secondary, the primary and the other secondary (assuming three-node replica sets) will continue as usual.

If it is a mongod process acting as a primary, then an election round will start to elect a new primary in this shard (which is really a replica set).

In both cases, we should actively monitor and try to repair the node as soon as possible, as our availability can be impacted.

Config server

Config servers must be configured as a replica set. A config server failing is no different than a regular mongod process failing. We should monitor, log, and repair the process.

A shard goes down

Losing an entire shard is pretty rare, and in many cases can be attributed to network partitioning rather than failing processes. When a shard goes down, all operations that would go to this shard will fail. We can (and should) implement fault tolerance at our application level, allowing our application to resume for operations that can be completed.

Choosing a shard key that can easily map on our operational side can also help; for example, if our shard key is based on location, we may lose the **European Union** (**EU**) shard, but will still be able to write and read data regarding **United States** (**US**)-based customers through our US shard.

The entire cluster goes down

If we lose the entire cluster, we can't do anything other than get it back up and running as soon as possible. It's important to have monitoring and to put a proper process in place to understand what needs to be done, when, and by whom, should this ever happen.

Recovering when the entire cluster goes down essentially involves restoring from backups and setting up new shards, which is complicated and will take time. Dry testing this in a staging environment is also advisable, as is investing in regular backups via MongoDB Ops Manager or any other backup solution.

> **Note**
> A member of each shard's replica set could be in a different location for **disaster-recovery** (**DR**) purposes.

Further reading

The following sources are recommended for you to study sharding in depth:

- *Scaling MongoDB* by *Kristina Chodorow*
- *MongoDB: The Definitive Guide* by *Kristina Chodorow* and *Michael Dirolf*
- MongoDB manual—Sharding: `https://docs.mongodb.com/manual/sharding/`
- MongoDB sharding pitfalls: `https://www.mongodb.com/blog/post/sharding-pitfalls-part-iii-chunk-balancing-and`
- MongoDB sharding and chunks: `http://plusnconsulting.com/post/mongodb-sharding-and-chunks/`
- MongoDB sharding internals: `https://github.com/mongodb/mongo/wiki/Sharding-Internals`
- MongoDB sharding: `http://learnmongodbthehardway.com/schema/sharding`

- MongoDB sharding and replica sets release (v1.6): `https://www.infoq.com/news/2010/08/MongoDB-1.6`

- Horizontal and vertical scaling explained: `http://www.pc-freak.net/images/horizontal-vs-vertical-scaling-vertical-and-horizontal-scaling-explained-diagram.png`

Summary

In this chapter, we explored sharding, one of the most interesting features of MongoDB. We started with an architectural overview of sharding and moved on to how we can design a shard and choose the right shard key.

We learned about monitoring, administration, and the limitations that come with sharding. We also learned about `mongos`, the MongoDB sharding router that directs our queries to the correct shard. Finally, we discussed recovery from common failure types in a MongoDB sharded environment.

The next chapter, on fault tolerance and high availability, will offer some useful tips and tricks that have not been covered in the other chapters.

15

Fault Tolerance and High Availability

In this chapter, we will try to fit in the information that we didn't manage to discuss in the previous chapters, and we will place emphasis on some other topics. Throughout the previous 14 chapters, we have gone all the way from covering the basic concepts of effective querying, to administration and data management, to scaling and **high-availability** (**HA**) concepts.

We will discuss how our application design should be accommodating and proactive with regard to our database needs. We will go over patterns and anti-patterns for schema design.

Day-to-day operations are another area that we will discuss, including tips and best practices that can help us to avoid nasty surprises down the line.

In light of the continued attempts by ransomware to infect and hold MongoDB servers hostage, we will offer more tips on security.

Finally, we will try to sum up the advice that's been given in a series of checklists that should be followed to ensure that the best practices are properly set up and followed.

This chapter covers the following topics:

- Application design
- Elevating operations
- Boosting security

Application design

In this section, we will describe some useful tips for application design that we did not cover or emphasize enough in the previous chapters.

Schema-less doesn't mean schema design-less

A big part of MongoDB's success can be attributed to the increased popularity of **object-relational maps (ORMs)/object document maps (ODMs)**. Especially with languages such as JavaScript and the **MongoDB, Express, Angular, and Node (MEAN)** stack, the developer can use JavaScript from the frontend (Angular/Express) to the backend (Node.js) to the database (MongoDB). This is frequently coupled with an ODM that abstracts away the internals of the database, mapping collections to Node.js models.

The major advantage is that developers don't need to fiddle with the database schema design, as this is automatically provided by the ODM. The downside is that database collections and schema designs are left up to the ODM, which does not have the business domain knowledge of different fields and access patterns.

In the case of MongoDB and other NoSQL-based databases, this boils down to making architectural decisions based not only on immediate needs but also on what needs to be done down the line. On an architectural level, this may mean that instead of a monolithic approach, we can combine different database technologies for our diverse and evolving needs by using a graph database for graph-related querying, a relational database for hierarchical, unbounded data, and MongoDB for **JavaScript Object Notation (JSON)** retrieval, processing, and storage.

In fact, many of MongoDB's successful use cases come from the fact that it's not being used as a one-size-fits-all solution, but only for the use cases that make sense.

Design patterns

Relational database management system (RDBMS) schema design has evolved and matured over the decades. The third normal form and **Boyce–Codd normal form (BCNF)** are primarily used to model data in a relational database.

MongoDB schema design has similarly evolved in the past decade, and there are several patterns and anti-patterns that can guide us when designing new databases or migrating existing workloads. We are going to examine both of these in the following sections.

Attribute pattern

The attribute pattern is commonly used when we have sparse attribute values across several logically grouped fields. We restructure a field/value pair into a subdocument with {key:field , val: value}.

For example, have a look at the following document with price information for a fan across the **United States (US)**, the **United Kingdom (UK)**, and Germany:

```
{
    Product: 'fan',
```

```
    Price_uk: '30',
    price_us: '35',
    Price_de: '40'
}
Would be restructured as:
{
    Product: 'fan',
    Prices: [
{    Country: 'uk', price: 30 },
{    Country: 'us', price: 35 },
{    Country: 'de', price: 40 }
]
}
```

This way, we can now create indexes for prices and countries and query them across all of our products.

Tree traversal pattern

Storing hierarchical, tree-like data is a common use case for any database. E-commerce classification of products by category and subcategory is probably the earliest application of hierarchical data storage in a database.

Data that follows a hierarchical structure can be stored in a traditional relational database with a standardized process using database and set theory, as explained in the bibliography and articles such as Mike Hillyer's blog post: `http://mikehillyer.com/articles/managing-hierarchical-data-in-mysql/`.

Storing a tree structure in a document-oriented database such as MongoDB is not much different. In this section, we will explain the fundamental building blocks that we can use to store and query our hierarchical data.

We have four different ways to traverse through a tree structure.

First of all, we can store a reference to the child document or documents. This can be as simple as storing the `objectId` value of the documents that are children to the current node.

Second, we can store a reference to the parent document. This can be as simple as storing the `objectId` value of the document that is parent to the current node.

These two methods are the simplest way to store parent-child references. Traversing the tree top down or bottom up can then be achieved by using the parent/child reference until there is none in the root and leaf nodes, respectively.

Third, we can store an array of ancestors for each document. Arrays are ordered by insertion order in MongoDB, so we need to insert the elements in order from the root to the leaf node. The last element is the parent node to the current document, and every element to the left of it in the array is its parent.

Finally, we can store the full list of ancestors in a string field, separated by the delimiter of our choice. This is similar to the third option, with the advantage that we can index and query using a **regular expression (regex)** for any category, from root to leaf.

For example, we may have a **fan** document under the following categorization:

```
home and kitchen -> kitchen and home appliances -> heating,
cooling and air quality -> fans -> Desk fans
```

Then, using the four methods here, we would have the following:

- `{ parent: 'heating, cooling and air quality' }`
- `{ children: ['Desk fans'] }`
- `{ ancestors: ['home and kitchen', 'kitchen and home appliances', 'heating, cooling and air quality'] }`
- `{ ancestors: 'homeandkitchen.kitchenandhomeappliances. heatingcoolingandairquality' }`

In all of these methods, we would be referencing the `ObjectId` values of the documents that belong to the given category.

Polymorphic pattern

Polymorphism in **object-oriented programming (OOP)** refers to the ability of a language to reuse the same interface for multiple heterogeneous underlying data types.

In MongoDB, polymorphism is omnipresent because of the flexibility it provides. Each document in a collection can have different fields and there is no restriction imposed by the database on the values that these fields may have.

This level of flexibility can result in increased maintenance costs for the application developers. We can restrict this flexibility by using JSON schemas, as we learned in previous chapters.

The polymorphic pattern attempts to keep the flexibility and deal with the complexity at the application level. We can use a field's value as a guide to indicate the structure of the document in our application.

For example, if we have a `'vehicles'` collection, we can have a field named `'type'` that can have a set of values such as `'car'`, `'bike'`, or `'motorcycle'`.

The application code can then be configured to expect different fields from a car (for example, mileage) than what it would expect from a bike. We can use **enumerators (enums)** or similar programming language structs to constrain the possible values of the type field.

Schema versioning pattern

MongoDB is schema-less, but as we have pointed out a few times already, we need to design our data storage in databases, collections, and documents. The schema versioning pattern addresses the need to change our schema over time.

It is one of the simplest patterns. All we need to do is add a field to declare the version of our schema. It can be, for example, `version` or `schema_version`. We start with `{version: 1}` and incrementally increase the version every time that we change the document schema by adding, modifying, or removing one or more fields. Our application code can adapt to the different document versions and use different business logic to process our data.

This pattern is useful when we need to change our schema without any downtime to adapt the structure of our documents to the new version.

Computed pattern

The computed pattern is used when we have many reads performing repeated calculations over a set of fields in a document. For example, if we have a user dashboard for a banking application that integrates with external services, we may have a counter of the number of active integrations. Instead of querying every time that we load the user profile for the subdocuments corresponding to the active integrations, we can calculate it at the time that we add or remove an integration and store it in a separate field.

Subset pattern

The subset pattern aims to strike a balance between the core MongoDB design direction that we should keep data that gets queried together in the same document, keeping unnecessary data in a single document.

We have already learned in previous chapters that we should not embed an unbounded number of fields or subdocuments in a single document because we will end up reaching the global document size limit of 16 **megabytes (MB)**.

We should be keeping data that gets queried together in the same document to avoid having to perform multiple reads across different collections to get our data.

What the subset pattern instructs us is that we should be trying to keep only the data that we need to query together in the same document. For example, if we have a document to store movie ratings for each user, we should only store the necessary metadata that we would display, along with each movie rating on the page, and store the extended metadata for each rating in a separate collection. This way, we can fetch all movie ratings in a single read and only fetch the extended metadata for each rating when the user clicks through to a different page to view the details of each rating.

This way, we can reduce our working set for the user collection so that it can easier fit into memory and avoid page swaps that would result in database slowdown.

Extended reference pattern

The extended reference pattern is similar and sometimes complementary to the subset pattern. Instead of separating our data fields based on where we need to access them, we are duplicating the data fields that we most need in order to avoid multiple queries across different collections.

Using the same example of users and movie ratings, we could duplicate the fields that we most frequently access from a movie rating into the users' collection to avoid querying the movie rating collection.

Similar to the subset pattern, we don't embed every single field but only the fields that we need to access at the same time as the user.

The downside of duplicating data is that we need to make sure that we update it in multiple places when we need to. Depending on the use case, this can be an advantage if we need to keep historical data with a snapshot at the time of creating the duplicating data. In that case, we would be updating the primary reference field and keeping copies of the data intact for archiving and reporting purposes.

Bucket pattern

The bucket pattern is most commonly used in **Internet of Things** (**IoT**) applications. For example, if we have a sensor reading values every minute, the most straightforward design is to store every reading in one document. This will quickly result in millions of tiny documents and, quite possibly, lots of data duplication.

We can instead use the bucket pattern and store one document for each extended time period—say, 1 day—reducing the number of documents to 1/1,440 minutes in a day. We can combine this with an index in the `sensor_id` field and archive older data as this becomes less probable to query in real time.

Outlier pattern

As with many of the other patterns discussed already, the outlier pattern is not unique to MongoDB. This pattern aims to address the commonly observed **Pareto principle of 80/20**.

Essentially, we should code for edge cases but not drive our design by outliers at the expense of a more efficient solution for the overwhelming majority.

For example, if we store movie ratings per user in an array, we may end up with very few users that have rated thousands of movies each. The vast majority of users are only rating a few movies each, and this is what should guide our design toward using an array to store the movie ratings.

We can code for the edge case of the few `'super-reviewers'` by adding an extra field in the document with a descriptive name such as `'super_reviewer'` and store the overflowing ratings in a separate collection.

Approximation pattern

More of a statistics trick than a pattern itself, the approximation pattern uses the law of large numbers to converge a randomly generated value with the actual value over time.

An example can be updating page views for a page. We can update page views by incrementing the `page_views` field by 1 each time a user visits the page. This will result in an accurate calculation of page views at any given time at the expense of having to update the document 1,000 times for every 1,000 page views.

We can also use `rand(0,999)` to generate a value between 0 and 999 inclusive and only update the counter by 1.000 when the value is 0, 1,337, or any other value of our preference. This will statistically only update the value every 1 in 1,000 times and over a long time will converge with the actual value.

This trick offloads a lot of the database workload at the expense of having an accurate page view count at any given time.

MongoDB published this pattern along with the rest of the patterns that we discussed in this section. More information is available in the link at the end of this chapter.

Design anti-patterns

MongoDB recommends avoiding the following design anti-patterns. Similar to the design patterns, these are dependent on the use case, and while we should generally avoid the practices outlined next, we should use our judgment according to our specific use case.

Massive arrays anti-pattern

Arrays are commonly used in MongoDB to store multiple values related to one field. Storing more than a few hundred values in one array can start having an impact in terms of performance, especially if we are using an index over that field. Storing an unbounded number of elements in an array should be avoided.

The recommended solution is to use the Extended Reference pattern and break down the array's elements into multiple parts.

Unnecessary indexes anti-pattern

Creating indexes is straightforward, and since it happens in the background, it will not affect our operations. Indexes speed up reads and by design slow down updates and inserts, as every update and insert needs to be reflected in the index as well. Indexes also take up space, which may become an issue when we have too many indexes.

The generic recommendation by MongoDB is to have fewer than 50 indexes per collection and to avoid wildcard indexes unless there is both a need and planning ahead for their needs.

We should design the indexes that we will use in our collections as much as we design the schema and data structures. The recommendations in *Chapter 8, Indexing,* should serve as guidance to design and implement the correct indexes over our data.

Bloated documents anti-pattern

One of MongoDB's core principles is to store in the same place data that is accessed together—that is, embed in the same document data that we will access together to avoid having to perform multiple queries.

Following this advice down to the T, we can end up with bloated documents that end up being several MB in size. This can be a problem because of the global document size limit of 16 MB.

Further to that, this slows down the database as every operation needs to fetch, update, and delete several MB instead of a few **kilobytes** (**KB**) at a time. The solution to this anti-pattern is to design ahead and only store in the same document data that is accessed together, resisting the temptation to store all our data in a single document and a single collection.

Case-insensitive queries without matching indexes anti-pattern

A common issue with indexes and queries is trying to query our data in a case-insensitive way when our indexes are case-sensitive. The index cannot be used and MongoDB resorts to a full collection scan, slowing down the query.

The solution to this problem is to create a case-insensitive index and/or to use a collation strength of 1 or 2 (the default is 3, which is case-sensitive) in the collection level.

Read performance optimization

In this section, we will discuss some tips for optimizing read performance. Read performance is directly correlated to the number of queries and their complexity. Performing fewer queries in a schema without complex nested data structures and arrays will generally result in better read performance. However, many times, optimizing for read performance can mean that the write performance will degrade. This is something to keep in mind and continuously measure when we are making performance optimizations in MongoDB.

Consolidating read querying

We should aim to have as few queries as possible. This can be achieved by embedding information into subdocuments instead of having separate entities. This can lead to an increased write load, as we have to keep the same data points in multiple documents and maintain their values everywhere when they change in one place.

The design considerations here are noted as follows:

- The read performance benefits from data duplication/denormalization
- The data integrity benefits from data references (DBRef or in-application code, using an attribute as a **foreign key (FK)**)

We should denormalize especially if our read/write ratio is too high (our data rarely changes values, but it gets accessed several times in between), if our data can afford to be inconsistent for brief periods of time, and—most importantly—if we absolutely need our reads to be as fast as possible and are willing to pay the price in consistency/write performance.

The most obvious candidates for fields that we should denormalize (embed) are dependent fields. If we have an attribute or a document structure that we don't plan to query on its own, but only as part of a contained attribute/document, then it makes sense to embed it, rather than have it in a separate document/collection.

Using our MongoDB books example, a book can have a related data structure that refers to a review from a reader of the book. If our most common use case is showing a book along with its associated reviews, then we can embed reviews into the book document.

The downside to this design is that when we want to find all of the book reviews by a user, this will be costly, as we will have to iterate all of the books for the associated reviews. Denormalizing users and embedding their reviews can be a solution to this problem.

A counterexample is data that can grow unbounded. In our example, embedding reviews along with heavy metadata can lead to an issue if we hit the 16 MB document size limit. A solution is to distinguish between data structures that we expect to grow rapidly and those that we don't and to keep an eye on their sizes through monitoring processes that query our live dataset at off-peak times and report on attributes that may pose a risk down the line.

> **Note**
> Don't embed data that can grow unbounded.

When we embed attributes, we have to decide whether we will use a subdocument or an enclosing array.

When we have a **unique identifier (UID)** to access the subdocument, we should embed it as a subdocument. If we don't know exactly how to access it or we need the flexibility to be able to query for an attribute's values, then we should embed it in an array.

For example, with our `books` collection, if we decide to embed reviews into each book document, we have the following two design options:

- Here's the code for a book document with an array:

```
{
Isbn: '1001',
Title: 'Mastering MongoDB',
Reviews: [
{ 'user_id': 1, text: 'great book', rating: 5 },
{ 'user_id': 2, text: 'not so bad book', rating: 3 },
]
}
```

- Here's the code for a book with an embedded document:

```
{
Isbn: '1001',
Title: 'Mastering MongoDB',
Reviews:
{ 'user_id': 1, text: 'great book', rating: 5 },
{ 'user_id': 2, text: 'not so bad book', rating: 3 },
}
```

The array structure has the advantage that we can directly query MongoDB for all of the reviews with a `rating` value greater than 4 through the embedded array reviews.

Using the embedded document structure, on the other hand, we can retrieve all of the reviews the same way that we would when using the array, but if we want to filter them, it has to be done on the application side, rather than on the database side.

Defensive coding

More of a generic principle, **defensive coding** refers to a set of practices and software designs that ensures the continuing functionality of a piece of software under unforeseen circumstances.

It prioritizes code quality, readability, and predictability. Readability was best explained by John F. Woods in his *comp.lang.c++* post, on September 24, 1991:

Always code as if the guy who ends up maintaining your code will be a violent psychopath who knows where you live. Code for readability.

Our code should be readable and understandable by humans, as well as by machines. With code quality metrics, as derived by static analysis tools, code reviews, and bugs reported/resolved, we can estimate the quality of our code base and aim for a certain threshold at each sprint, or when we are ready to release. Code predictability, on the other hand, means we should always expect results in unexpected input and program states.

These principles apply to every software system. In the context of system programming using MongoDB, there are some extra steps that we must take to ensure that code predictability and—subsequently—quality are measured by the number of resulting bugs.

MongoDB limitations that will result in a loss of database functionality should be monitored and evaluated on a periodic basis, as follows:

- **Document size limit**: We should keep an eye on collections in which we expect to have documents growing the most, running a background script to examine document sizes and alert us if we have documents approaching the limit (16 MB), or if the average size has grown significantly since the last check.

- **Data integrity checks**: If we are using denormalization for read optimization, then it's a good practice to check for data integrity. Through a software bug or a database error, we may end up with inconsistent duplicate data among collections.

- **Schema checks**: If we don't want to use the document validation feature of MongoDB, but rather we want a lax document schema, it's still a good idea to periodically run scripts to identify fields that are present in our documents and their frequencies. Then, along with relative access patterns, we can identify whether these fields can be identified and consolidated. This is mostly useful if we are ingesting data from another system wherein data input changes over time, which may result in a wildly varying document structure at our end.

- **Data storage checks**: This mostly applies when using **Memory Mapped Storage Engine version 1 (MMAPv1)**, where document padding optimization can help performance. By keeping an eye on document size relative to its padding, we can make sure that our size-modifying updates won't incur a move of the document in physical storage.

These are the basic checks that we should implement when defensively coding for our MongoDB application. On top of this, we need to defensively code our application-level code to make sure that when failures occur in MongoDB, our application will continue operating—perhaps with degraded performance, but still operational.

An example of this is replica set failover and failback. When our replica set primary fails, there is a brief period to detect this failure and the new primary is elected, promoted, and operational. During this brief period, we should make sure that our application continues to operate in read-only mode, instead of throwing a 500 error code. In most cases, electing a new primary is done in seconds, but in some cases, we may end up in the minority end of a network partition and unable to contact a primary for a long period of time. Similarly, some secondaries may end up in a recovering state (for example, if they fall way behind the primary in replication); our application should be able to pick a different secondary in this case.

Designing for secondary access is one of the most useful examples of defensive coding. Our application should weigh between fields that can only be accessed by the primary to ensure data consistency and fields that are okay to be updated in near real time, instead of in real time, in which case we can read these from secondary servers. By keeping track of replication lag for our secondaries by using automated scripts, we can have a view of our cluster's load and how safe it is to enable this functionality.

Another defensive coding practice is to always perform writes with **Journaling** on. Journaling helps to recover from server crashes and power failures.

Finally, we should aim to use replica sets as early as possible. Other than the performance and workload improvements, they help us to recover from server failures.

Monitoring integrations

All of this adds up to greater adoption of monitoring tools and services. As much as we can script some of them, integrating with cloud and on-premises monitoring tools can help us to achieve more in a smaller amount of time.

The metrics that we keep a track of should do one of the following:

- **Detect failures**: Failure detection is a reactive process, where we should have clear protocols in place for what happens when each of the failure-detection flags goes off. For example, what should the recovery steps be if we lose a server, a replica set, or a shard?

- **Prevent failures**: Failure prevention, on the other hand, is a proactive process, designed to help us catch problems before they become a potential source of failure in the future. For example, **central processing unit** (**CPU**)/storage/memory usage should be actively monitored with yellow and red thresholds, and clear processes should be put in place as to what we should do in the event that we reach either threshold.

Application design and best practices can help us get ahead of the game. Both are a continuous journey that we need to embark on from the greenfield stage all the way to maintenance and eventual decommissioning of the project. In the next section, we will focus on a few operational tips that we can use to be proactive and foresee any potential performance issues.

Elevating operations

When connecting to our production MongoDB servers, we want to make sure that our operations are as lightweight as possible (and are certainly non-destructive) and do not alter the database state in any sense.

Two useful utilities that we can chain to our queries are shown here:

```
> db.collection.find(query).maxTimeMS(999)
```

Our `query` instance will only take up to 999 **milliseconds (ms)**, and will then return an exceeded time limit error, as follows:

```
> db.collection.find(query).maxScan(1000)
```

Our `query` instance will examine 1000 documents at the most, in order to find results and then return (no error raised).

Whenever we can, we should bind our queries by time or document result size to avoid running unexpectedly long queries that may affect our production database. A common reason for accessing our production database is troubleshooting degraded cluster performance. This can be investigated via cloud monitoring tools, as we described in previous chapters.

The `db.currentOp()` command, through the MongoDB shell, will give us a list of all current operations. We can then isolate the ones that have large `.secs_running` values and identify them through the `.query` field.

If we want to kill an in-progress operation that takes a long time, we need to note the value of the `.opid` field and pass it on `db.killOp(<opid>)`.

Finally, it's important to recognize (from an operational standpoint) that everything may go wrong. We must have a backup strategy in place that is implemented consistently. Most importantly, we should practice restoring from backup to make sure that it works as intended.

We should be extremely cautious when issuing any command on a live cluster and dry run all operations in a testing/staging environment that mirrors the production environment as closely as possible. The last part of this chapter will provide a few tips to boost security from a holistic perspective.

Boosting security

After the recent waves of ransomware that were locking down unsecured MongoDB servers, asking for ransom payments in cryptocurrency from the administrators to unlock the MongoDB servers, many developers have become more security-conscious. Security is one of the items on a checklist that we, as developers, may not prioritize highly enough, due to the optimistic belief that it won't happen to us. The truth is, in the modern internet landscape, everyone can be a target of automated or directed attacks, so security should always be taken into account, from the early stages of the design to after production deployment.

Enabling security by default

Every database (other than local development servers, perhaps) should be set up with the following in the `mongod.conf` file:

```
security:
    authorization: enabled
```

> **Note**
>
> **Secure Sockets Layer (SSL)** should always be enabled, as we described in the relevant chapter: *Chapter 9, Monitoring, Backup, and Security*.

Access should be restricted to only allow communication between application servers and MongoDB servers, and only in the interfaces that are required. Using `bind_ip`, we can force MongoDB to listen to specific interfaces, instead of the default binding to localhost behavior, as follows:

```
net:
    bindIp: 127.0.0.1, 10.10.0.10,10.10.0.20
```

Isolating our servers

We should secure our infrastructure perimeter with **Amazon Web Services (AWS) Virtual Private Cloud (VPC)** or the equivalent from the cloud provider of our choice. As an extra layer of security, we should isolate our servers in a cloud of their own, only allowing external connections to reach our application servers and never allowing them to directly connect to our MongoDB servers, as illustrated in the following diagram:

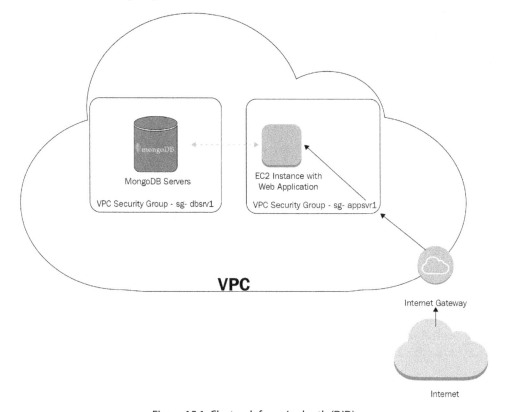

Figure 15.1: Cluster defense in depth (DiD)

We should invest in role-based authorization. Security lies not only in protecting against data leaks caused by external actors but also in making sure that internal actors have the appropriate levels of access to our data. Using role-based authorization at the MongoDB level, we can make sure that our users have the appropriate levels of access.

Consider Enterprise Edition for large deployments. Enterprise Edition offers some convenient features concerning security and more integrations with well-known tools and should be evaluated for large deployments, with an eye on changing needs as we transition from a single replica set to an enterprise-complex architecture.

Checklists

Operations require the completion of many tasks and complexity. A good practice is to keep a set of checklists with all of the tasks that need to be performed and their order of significance. This will ensure that we don't let something slip through. A deployment and security checklist, for example, could look like this:

- **Hardware**:

  - **Storage**: How much disk space is needed per node? What is the growth rate?

  - **Storage technology**: Do we need a **solid-state drive** (SSD) versus a **hard disk drive** (HDD)? What is the throughput of our storage?

  - **RAM**: What is the expected working set? Can we fit it in the **random-access memory** (RAM)? If not, are we going to be okay with an SSD instead of an HDD? What is the growth rate?

  - **CPU**: This usually isn't a concern for MongoDB, but it could be if we planned to run CPU-intensive jobs in our cluster (for example, aggregation or `MapReduce`).

  - **Network**: What are the network links between servers? This is usually trivial if we are using a single data center, but it can get complicated if we have multiple data centers and/or offsite servers for **disaster recovery** (DR).

- **Security**:

  - Enable auth.

  - Enable SSL.

  - Disable **REpresentational State Transfer** (REST)/**Hypertext Transfer Protocol** (HTTP) interfaces.

  - Isolate our servers (for example, VPC).

  - Authorization is enabled. With great power comes great responsibility. Make sure that the powerful users are the ones that you trust. Don't give potentially destructive powers to inexperienced users.

A monitoring and operations checklist could look like this:

- **Monitoring**:

  - Usage of hardware (CPU, memory, storage, and network).

  - Health checks, using Pingdom or an equivalent service to make sure that we get a notification when one of our servers fails.

  - **Client performance monitoring**: Integrating periodic mystery shopper tests using the service as a customer in a manual or automated way, from an **end-to-end** (**E2E**) perspective, in order to find out whether it behaves as expected. We don't want to learn about application performance issues from our customers.

  - Use MongoDB Cloud Manager monitoring; it has a free tier, it can provide useful metrics, and it is the tool that MongoDB engineers can take a look at if we run into issues and need their help, especially as a part of support contracts.

- **DR**:

  - **Evaluate the risk**: What is the risk, from a business perspective, of losing MongoDB data? Can we recreate this dataset? If yes, how costly is it in terms of time and effort?

  - **Devise a plan**: Have a plan for each failure scenario, with the exact steps that we need to take if something happens.

  - **Test the plan**: Having a dry run of every recovery strategy is as important as having one. Many things can go wrong in DR, and having an incomplete plan (or one that fails in each purpose) is something that we shouldn't allow to happen in any circumstance.

  - **Have an alternative to the plan**: No matter how well we devise a plan and test it, anything can go wrong during planning, testing, or execution. We need to have a backup plan for our plan, in case we can't recover our data using plan A. This is also called plan B, or the last-resort plan. It doesn't have to be efficient, but it should alleviate any business reputation risks.

  - **Load test**: We should make sure that we load test our application E2E before deployment, with a realistic workload. This is the only way to ensure that our application will behave as expected.

Summary

In this chapter, we covered some topics that were not detailed in previous chapters. It is important to apply the best practices according to our workload requirements. We started by covering some patterns and anti-patterns that MongoDB has identified over the years. Read performance is usually what we want to optimize for; that is why we discussed consolidating queries and the denormalization of our data.

Operations are also important when we go from deployment to ensuring the continuous performance and availability of our cluster. Security is something that we often don't think about until it affects us. That's why we should invest the time beforehand to plan and make sure that we have the measures in place to be sufficiently secure.

Finally, we introduced the concept of checklists to keep track of our tasks and to make sure that we complete all of them before major operational events (deployment, cluster upgrades, moving to sharding from replica sets, and so on)

Further reading

You can refer to the following links for further information:

- MongoDB building with patterns: `https://www.mongodb.com/blog/post/building-with-patterns-a-summary`
- MongoDB University *Data Modeling* course: `https://university.mongodb.com/courses/M320/about`
- MongoDB tips and tricks: `http://mo.github.io/2017/01/22/mongo-db-tips-and-tricks.html`
- SQL to MongoDB: `https://studio3t.com/whats-new/tips-for-sql-users-new-to-mongodb/`
- MongoDB performance tips: `https://www.hostreview.com/blog/170327-top-7-mongodb-performance-tips-must-know`
- Code for readability (Woods): `https://groups.google.com/forum/#!msg/comp.lang.c++/rYCO5yn4lXw/oITtSkZOtoUJ`

Closing remarks

Congratulations for making it this far and thank you for embarking on this journey with me for the past 15 chapters! It's been 11 years since I first visited MongoDB's first office ever in New York to attend a multi-day training. As a young engineer, I was inspired not only by their technical merits but equally as much by their behavioral traits, their friendly and hardworking attitude, and overall professional ethics.

Being an outsider and still very close to the team for a few days, I could clearly see that this team would succeed. In these 11 years, a lot has changed in my life. My beloved mother, Evi, and my dearest father-in-law, Christos, passed away and a new member of our family is now on the way.

On the other hand, MongoDB has been a constant source of satisfaction and joy to work with. I was lucky enough to witness firsthand its whole evolution from a small, niche database to a fully fledged ecosystem that can serve almost everyone, from solo founders to the major *Fortune 500* corporations.

If there is one thing I learned watching MongoDB's story unfold, it is that small gains add up over time. We can't become the best version of ourselves in a day, but hard work, ethics, and persistence always pay off over time. Getting just a little bit better every day as a professional and as a member of society is at least one of the ways to ensure a better future for everyone.

In the words of Dale Carnegie: "*Most of the important things in the world have been accomplished by people who have kept on trying when there seemed to be no hope at all.*"

Live long and prosper!

Index

Y

Z

`Packt.com`

Subscribe to our online digital library for full access to over 7,000 books and videos, as well as industry leading tools to help you plan your personal development and advance your career. For more information, please visit our website.

Why subscribe?

- Spend less time learning and more time coding with practical eBooks and Videos from over 4,000 industry professionals

- Improve your learning with Skill Plans built especially for you

- Get a free eBook or video every month

- Fully searchable for easy access to vital information

- Copy and paste, print, and bookmark content

Did you know that Packt offers eBook versions of every book published, with PDF and ePub files available? You can upgrade to the eBook version at `packt.com` and as a print book customer, you are entitled to a discount on the eBook copy. Get in touch with us at `customercare@packtpub.com` for more details.

At `www.packt.com`, you can also read a collection of free technical articles, sign up for a range of free newsletters, and receive exclusive discounts and offers on Packt books and eBooks.

Other Books You May Enjoy

If you enjoyed this book, you may be interested in these other books by Packt:

MongoDB Fundamentals

Amit Phaltankar , Juned Ahsan , Michael Harrison, Liviu Nedov

ISBN: 978-1-83921-064-8

- Set up and use MongoDB Atlas on the cloud
- Insert, update, delete, and retrieve data from MongoDB
- Build aggregation pipelines to perform complex queries
- Optimize queries using indexes
- Monitor databases and manage user authorization
- Improve scalability and performance with sharding clusters
- Replicate clusters, back up your database, and restore data

Packt is searching for authors like you

If you're interested in becoming an author for Packt, please visit `authors.packtpub.com` and apply today. We have worked with thousands of developers and tech professionals, just like you, to help them share their insight with the global tech community. You can make a general application, apply for a specific hot topic that we are recruiting an author for, or submit your own idea.

Share Your Thoughts

Now you've finished *Mastering MongoDB 6.x, Third Edition,* we'd love to hear your thoughts! Scan the QR code below to go straight to the Amazon review page for this book and share your feedback or leave a review on the site that you purchased it from.

https://packt.link/r/1-803-24386-4

Your review is important to us and the tech community and will help us make sure we're delivering excellent quality content.

www.ingramcontent.com/pod-product-compliance
Lightning Source LLC
Chambersburg PA
CBHW081458050326
40690CB00015B/2839